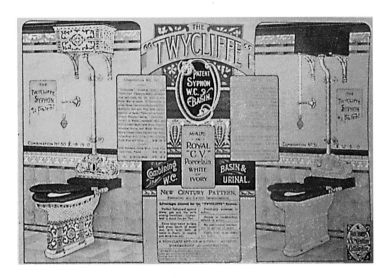

TEMPLES OF CONVENIENCE
AND
CHAMBERS OF DELIGHT

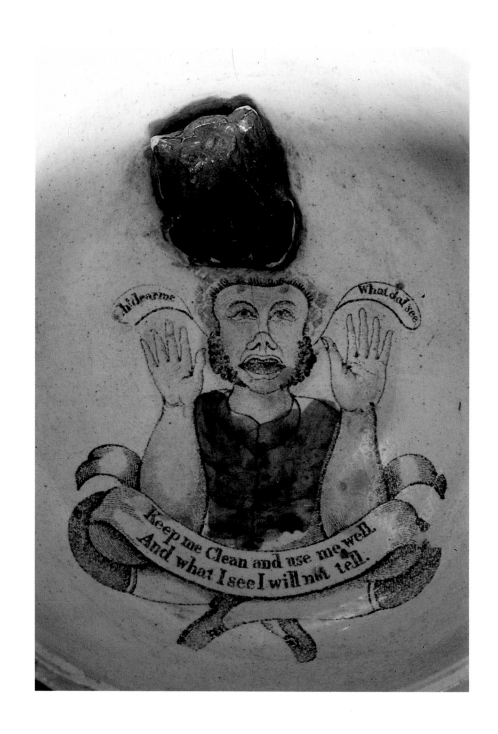

TEMPLES OF CONVENIENCE
AND
CHAMBERS OF DELIGHT

LUCINDA LAMBTON

PAVILION

To Paul Keegan and to
Justine Fulford
with love and thanks

This edition published in Great Britain in 1997 by
Pavilion Books Limited
26 Upper Ground
London SE1 9FD

Originally published in hardback in 1995

Introduction and captions copyright © 1978 by Lucinda Lambton
This edition 1995
The moral right of the author has been asserted
Photographs copyright © 1995 by Lucinda Lambton except:
BBC Hulton Picture Library, page 11
Mary Evans Picture Library, pages 7, 13, 14, 18, 19, 21, 22
The Science Museum/Science and Society Picture Library, pages 12, 25

Designed by Andrew Barron & Collis Clements Associates

A CIP catalogue record for this book is available from the British Library

ISBN 1 85793 915 8

Printed and bound in Slovenia by Mladinska Knjiga

2 4 6 8 10 9 7 5 3 1

This book may be ordered by post direct from the publisher.
Please contact the Marketing Department.
But try your bookshop first.

Illustration on page 2: The Duke of Wellington's pot at Stratfield Saye, Berkshire.
Page 3: Lord Bute's bathroom of polished marble at Cardiff Castle.

CONTENTS

INTRODUCTION

It would have a good effect upon men to realize that the very vessels which
they make use of for their most dishonourable and base purposes, the receiving
of their dung and excrements, are made of as good, nay the same materials as
their own bodies – the finest and most beautiful bodies are but earthen vessels as
well as chamber pots; they are but statues made of clay, and are therefore as brit-
tle as their chamber pots and close stool pans, and will certainly be reduced and
crumbled into as many pieces.

THE PHILOSOPHICAL DIALOGUE CONCERNING DECENCY, 1851

Hail to the lavatory! An exceedingly important feature of
our lives.

The first – a flushing water-closet, with working parts – was
invented in 1596 by Sir John Harington, the godson of Queen
Elizabeth I. His apparatus had a brass valve to release the contents of the 'stool pot', which was made of either stone, lead or
brick; and a cistern, which in his diagram is filled with water,
represented by little fish swimming through the ripples. There
was a waste-pipe, too, as well as a washer, and a 'stopple' and a
great brass sluice through which everything flowed, 'down a gal-
lop into the jax'. 'This devise of mine', wrote Harington, in his
Metamorphosis of Ajax (a learned and elaborate satire on the
water-closet) requires not a sea full of water, but a cisterne, not
a whole Tems [Thames] full, but halfe a tunne full, to keep all
sweete and savourie . . . Reviced with water as oft as occasion
serves, but especially at noone and at night, will keep your priv-
ie as clean as your parlour and perhaps sweeter too.' He installed
one in his house at Kelstone near Bath and made one for his
godmother, the Queen, at Richmond Palace, but his enticing
schemes were to no further avail. He was 179 years before his
time – although James I granted a patent for a water-closet in
1617, nothing more was heard of it – and the reeking, stinking
years were to go on until 1775, when the first wholly successful
flushing water-closet was created to cleanse the kingdom.

The Romans, of course, had established an extraordinarily
sophisticated – as well as elegant and efficient – sanitary system
in Britain. From the splendour of Bath (Aquae Sulis) down to
the dextrous construction of a pipe, one finds an understanding
and appreciation of sanitary needs that were not to appear again,
on such a scale, for more than 1,500 years. Their baths were
steeped in luxury, with pools of hot and cold water as well as
sweating rooms and showers. Wallowing in splendour, the bather
was spread with unguents and oils, or sand if he was especially
dirty, then scraped clean with a metal 'strigil'. There were quantities of public baths, so there was little need to have one in your
house – although there was an immense private bathing pool, 36
feet by 18 feet, at a Roman villa in Chipping Warden,
Northamp-tonshire. Less appealingly, the latrines too were public, as at Housesteads on the Roman Wall in Northumberland,
where as many as twenty men would sit together, enjoying the
sight and company of their companions sending their offerings
to Stercutius (Saturn) and Crepitus, the gods of 'odure' and conveniences, as well as to Cloacina, the goddess of the common
sewer. Cloacina was one of the first of the Roman deities,
believed to have been named by Romulus himself. She was the
purifier who cleansed all, having under her charge every drain

and sewer (*cloaca*). Stercutius was one of Saturn's surnames, given to him when he laid dung upon the earth to make it fertile.

> The Romans, ever counted superstitious,
> Adored with high titles of divinity,
> Dame Cloacina and the Lord Stercutius –
> Two persons, in their state, of great affinity.
>
> SIR JOHN HARINGTON, *METAMORPHOSIS OF AJAX*

The Romans painted the walls of their latrines with deities and other hallowed emblems, to protect them against 'such as commit nuisance', and announced 'the wrath of Heaven against those who should be impious enough to pollute what it was their duty to reverence' (The Revd John James Blunt, *Vestiges of Ancient Manners and Customs*). Vessels for relief were put on street corners and could be enjoyed without charge, until Vespasian (who had rented them out to all those who would profit from the urine) put a tax on their use, having seen the valuable liquid flowing so freely. These smaller containers would eventually be emptied into larger urinary reservoirs, and used for the fulling of cloth. Pliny noticed that the Roman fullers who used human urine never suffered from gout.

There were no latrines or privies attached to houses. Basins and pots were used, looked after by slaves employed exclusively for the tempting tasks. 'When a gentleman wanted his chamber-pot,' wrote Petronius, '. . . it was a common way of speaking to make a noise with the finger and thumb, by snapping them together. This was called "concrepare digitos".'

A most miserable fate awaited many who walked beneath windows at night, when the contents of these pots were thrown down into the streets below – 'Clattering the storm descends from heights unknown' (*Third Satire* of Juvenal).

THEIR BATHS WERE STEEPED IN LUXURY, WITH POOLS OF HOT AND COLD WATER AS WELL AS SWEATING ROOMS AND SHOWERS.

Dryden further translated the grim possibilities:

> 'Tis want of sense to sup abroad too late
> Unless thou first has settled thy estate;
> As many fates attend thy steps to meet
> As there are waking windows in the street;
> Bless the good Gods and think thy chance is rare
> To have a piss-pot only for thy share.

There was no lavatory paper in the public latrines; instead sponge sticks were kept in containers of salt water, or dipped into running water, and used by all. The poorer Roman might use a stone for cleansing, or perhaps a shell or a bunch of herbs. Seneca wrote of the horror of a German slave who committed suicide by ramming such a sponge stick down his throat. There was a grim tale, too, of Roman soldiers who, thinking themselves disgraced by being asked to build a common sewer, had all committed suicide together. Latrines, privies and drains were therefore not always bathed in the light of the deities, although Cloacina was worshipped: 'Goddess of the stools, the jakes and the privy to whom, as to every one of the rest, there was a peculiar temple edified' (Reginald Scott, *The Discovery of Witchcraft*). Titus Tacius – who ruled with Romulus – had a statue of Cloacina in his magnificent privy, which had been built as a fitting shrine to her glory.

The luxuriant depravity of the Romans reached its peak with chamber-pots, and other such vessels, being made of rare stones and metals. 'It would have been well for the Romans', wrote Rolleston in 1751, 'if they had but remained contented with earthen jurdens – we may date the commencement of [their] ruin from the introduction of gold and silver chamber-pots and close-stool pans.' He went on to write of the Emperor

Heliogabalus, 'a monster of lust, luxury and extravagance' who, according to Lampridus, 'owned close-stool pans of gold, but his chamber pots were made, some of myrrh and stones of onyx'. The Revd Nathaniel Waverly, in his *Wonders of The Little World*, tells us more of the Emperor: 'His excrements he discharged into gold vessels and urined into vessels of onyx, or myrrhine pots . . . and even these, and the most part of his other vessels had lascivious engravings represented on the sides of them.' Heliogabalus's end was just and fitting: he was killed in a public latrine.

'The chamber pot had been invented by the Sybarites because they would not be at the trouble of moving'. So wrote the Revd Thomas Dudley Fosbroke in his *Encyclopedia of Antiquities* in 1825.

To make such a dicovery, in such an obscure book, was like coming upon the inventor of the wheel. How could it be that it is known who created the first chamber pot? With the help of a classicist friend, Rowland Smith, I ran the source to ground in The Bodleian Library in Oxford, and will never forget the words as they were crisply and clearly translated from their original Greek:

The horsemen of the Sybarites, more than five thousand strong, paraded with saffron-coloured coats over their breast plates, and in the summer their young men journeyed to the grottoes of the nymphs on the Lusias river and there spent the time in every form of luxury. Whenever the wealthy among them went for a vacation to the country they took three days to finish the one-day journey, although they travelled in carriages. Further, some of their roads leading to the country were roofed over. Most of them own wine cellars near the sea shore, into which the wines are sent through pipes from their country estates...They also hold many public banquets at frequent intervals, and they reward with golden crowns the men who have striven brilliantly for honours, and publish their values at the state sacrifices and games, proclaiming not so much their loyalty to the state as their service in providing dinners; on these occasions they crown even their cooks who have skilfully concocted the dishes served.

Among the Sybarites were also devised tubs in which they lay and enjoyed vapour baths. They too, were the first to invent chamber pots, which they carried to their drinking parties'.

Eureka! For anyone saturated with the history of sanitation, it was like coming upon the very cradle of civilization.

Another person to emerge from the shadowy ages of sanitary history was Boniface – writer of the first Latin grammar in England – when he spoke out against mixed bathing in AD 745. It was the social licentiousness of it that he was decrying, rather than the cleansing of the body, but to the early Christian church washing had become a most viceful activity, and this cultivation of the body was a luxury that had to be denied. Saint Francis of Assisi considered water to be one of the most precious of 'sisters' and denied himself of it entirely. Saint Catherine of Siena, as well as never washing, succeeded with the ultimate self-denial – refusing to relieve herself for months on end.

Conversely and confusingly, the use of water was a most holy baptismal and purifying rite, and by the twelfth century religious houses were employing it freely to establish their healthy and holy lives. Physical cleanliness was a sign of spiritual cleanliness. A very early account exists of Ethelwold, Abbot of Abingdon, who in AD 960 is recorded to have made a watercourse (*ductum aquae*), which ran under the dormitory to the 'Hokke' stream. This was obviously an efficient diversion of a stream or spring, and its description is the first detailed post-Roman account of sanitation in Britain.

The next, more elaborate development was carried out between 1150 and 1175, when a complete and complex drainage system was laid down in the Christ Church Monastery and the Cathedral Priory at Canterbury. The Monastery at Durham, too, had an elaborate watercourse, with an immense and magnificent 'fair laver or conduit, for the monks to wash their hands and faces at, being made in form round, and covered with lead, and all of marble, saving the very outermost walls: within which walls you may walk round about the laver of marble, having made many little conduits or spouts of brass within XXIII cockles of brass round about it, having in it seven fair windows, of stone work' (*The Ancient Rites of Durham*, quoted by Willis). This must have been very like the beautiful circular laver at Much Wenlock Priory in Shropshire, part of which survives to this day – with a wealth of twelfth-century carvings. The privies at Durham are also described in *Ancient Rites*: 'There was a fair large house and a most decent place adjoining to the West side of the said dorter [dormitory] towards the water, for the monks and the novices to resort unto, called the privies . . . and every seat and partition was of wainscot close, on either side very decent, so that one of them could not see one another, when they were in that place. There were as many seats of privies on either side as there were little windows in the walls, which windows were to give light to everyone on the said seats.'

The holy orders set an example to all: at Furness Abbey in Lancashire the seats of the necessarium were built back to back in a long row, and at Cleeve Abbey in Somerset the River Washford was diverted to flow beneath the privies. An enormous reredorter was built at Fountains Abbey in Yorkshire, with twenty-seven privies on two storeys. The Cistercian abbey at Waverly had a water supply by 1179, while the religious houses of Whalley, Winchester and Chester all had water, and the convent at Chester was supplied with water from an enormous tank at Christleton, some 3 miles away. At Tintern the tidal River Severn was ingeniously used to flush the reredorter at intervals. An extraordinarily beautiful lavatorium was built in Gloucester, parallel to the great fan-vaulted Cathedral cloisters, in a miniature fan-vaulted passage of its own. In 1924 excavations were carried out in the great cloister of St Albans Cathedral and, according to H.A.J. Lamb, writing in *The Architects' Journal* in 1937:

> Here was found a deep pit, 18 feet 8 ins long, by 5 feet 3 ins wide; the walls were 15 feet thick. The depth of the pit below the cloister floor was 25 feet. At the bottom were found pieces of pottery and fragments of coarse cloth, which, it is thought, were old gowns torn up by the monks and used as toilet paper. Evidence, too, that the monks suffered from digestive troubles, which were by no means rare in those days, was proved by finding in the pit, seeds of the blackthorn – a powerful aperient.

London got its first public water supply in 1237, when King Henry III requested that Gilbert de Sandford should grant to the city the springs and waters of his land at Tyburn. The first major sanitary act to be passed in London was achieved in 1358, when the 'Chancellor of The University' was required by royal writ 'to remove from the streets and lanes of the town, all swine and all dirt, dung, filth . . . and to cause [them] to be kept clean for the future'. There had been an earlier act of 1189 'concerning the necessary chambers in the houses of the citizens'. It stipulated that if a cesspit was lined with stone it should be a minimum of $2^1/_2$ feet from the neighbours' land, and if not so lined it should be a foot further away. This had subsequently proved to be a miserable necessity, as was shown in 1328 when a William Sprot complained to the Assize about his neighbours, brothers William and Adam Mere, who had filled their 'cloaca' to overflowing, so that its contents had seeped through the wall. It had not been the required distance away. At another Assize court, in 1347, two men were accused of piping their 'odure' into their neighbour's cellar, which villainous act had not been discovered until the cellar overflowed into the house.

Cesspits were an alternative to the open drains and sewers, which flowed into the turgid rivers. In 1355 the River Fleet had been discovered to be choked solid with filth from the eleven overhanging latrines and the three sewers that disgorged themselves into it. By this time there had been an unwise relaxing of the law forbidding overhanging privies. They had been illegal for years, but this was largely ignored by those who could afford them, so in 1383 the city authorities finally gave way and allowed projecting garderobes (privies) to be built, providing, of course, that the house abutted water, and providing also that no one 'throw rubbish or other refuse through the same'. Each household that built such a latrine was to pay a yearly revenue of two shillings (10p) to the Lord Chamberlain, towards keeping their watercourse clear.

There were three main watercourses in London: the River Fleet and the Walbrook stream, which both ran into the Thames. As most of the public latrines were built over the rivers, they became grotesquely overburdened by the ever-increasing volume of filth flowing into them, and by 1462 the latrine law had to be revoked. Orders were issued for Walbrook to be paved over and all the latrines to be removed; no more privies could be built over the Fleet. The Thames went on being as repulsive as ever; to go under London Bridge was thought to be the act of a fool, to go over it that of a wise man. All three watercourses were foul and disease-ridden veins that flowed through the city; astonishingly, they were to remain so until 1858, when Sir Joseph Bazalgette began to lay his great underground sewerage system, with Doulton stoneware pipes, beneath London.

Pots were the unpleasant alternative to latrines, their contents simply being thrown down on to the streets and the passers-by below. The open drains were clogged with foul rubbish, as well as putrid entrails and dung and dirt. Conditions became so bad that the Common Council appointed 'scavengers', both to clean town streets and to impose fines on those who added to the dirt. Shakespeare's father was so fined, for leaving refuse in the streets and for failing to keep his gutters clean; and the Rector of St Botolph's had to appear before the Assize of Nuisances for having allowed offensive piles of filth to accumulate round his privy. By 1345 the fine had risen to two shillings (10p) from its original sum of four pence (2p) and a good deal of effort was made to avoid having to pay it: a pedlar who threw eel skins on to the street was set upon by enraged citizens, who actually killed him in their fury.

The privies and latrines, both public and private, also had to be cleaned; a cesspit with the accumulated filth of months was an atrocious prospect for the 'rakers' and 'gongfermors' as they were called ('gong' from the Saxon word *gang,* to go off, which is still in use in the north-east of England today; 'fer' from the Saxon verb *fey,* to cleanse). Their pay, though, was exceedingly handsome: some forty shillings (£2) a job. One such, known as Richard the Raker, met with a dreadful death in his own privy in 1326, when he fell through the rotten planks, and drowned 'monstrously in his own excrement' (*The Black Death*). According to the tale of the fate of the 'Jew of Tewkesbury', in 1259, was even grimmer: he fell into a privy pit one Saturday and, out of respect for his Sabbath, no one was permitted to pull him out. On Sunday, thanks to the mysterious intervention of the Bishop of Gloucester, no one was allowed to rescue him, and

by Monday the poor fellow was dead (*Scatological Rites*).

Westminster Palace was the first lay house in the British Isles to have an underground drainage system, having had a water supply from as early as 1233. Henry III, who installed it, appears to have been a veritable saint of sanitation by contemporary standards. He built privies and garderobes – as well as 'wardrobes', 'privy chambers' and 'necessary places' – into all his palaces, insisting on them wherever he went. He even went so far as having them built in houses that he was visiting. At Clipstone his garderobes were a serious architectural addition, with shingled walls and glass windows, while at Winchester he ordered that the Queen's chamber was to have a cistern 'and likewise a chimney in it and to paint a certain city over the door of that chamber' (*Domestic Architecture in the Middle Ages*).

Privy is an Early Middle English word that comes from the latin *privatus,* meaning 'apart, retired, secret, not publicly known'. The word garderobe is sometimes confused with wardrobe. The wardrobe was the room adjoining the privy, in which you might wash and dress. Clothes were kept in it, and sewing and dressmaking might also be done there. Garderobe was the usual word for the privy in the grander domestic dwelling or the castle, where it might be built into the thickness of the wall and approached by a right-angled passage – a kind of horizontal trap to waylay the wafting odours. Garderobes could alternatively be corbelled, or simply built out from the walls, allowing everything to fall freely down to the moat, river or stream below. Sometimes there was a pit, but others just depended on the ground below. At Beaumaris in Wales the filth was expelled through the mouths of grimacing stone faces, but at Haddon Hall in Derbyshire there was just a straight drop on to a natural slant of sheer rock. These methods were often unsuccessful, as stained walls beneath these little projections still testify today. Chutes would descend from privies that were within the building, although at Broughton Castle in Oxfordshire there was a covered-in chute that was built on to the outer wall, beneath the corbelling. In the Palace of the Archbishop of York at Southwell in Nottinghamshire there is a simple pit beneath a discreet semicircular arrangement of seats.

Edward II, according to Marlowe, was kept in such a garderobe pit at Berkeley Castle in Gloucestershire, while his Queen, Isabella, and the nobles decided on his fate. In the last act, Sir John Matrevis, wondering at the King's strength, is talking to Sir John Gurney:

MATREVIS Gurney, I wonder the King dies not,
 Being in a vault up to the knees in water,
 To which the chambers of the castle run,
 From whence a damp continually ariseth
 That were enough to poison any man,
 Much more a King, brought up so tenderly.
GURNEY And so do I Matrevis. Yester night I open'd but the door to throw him meat, and I was almost stifled by the savour.

Edward, later to be done to death by Lightbourne, describes his ordeal to him:

EDWARD This dungeon where they keep me is the sink
 Wherein the filth of all the castle falls[...]
 And there, in mire and puddle have I stood
 This ten days' space; and, lest I should sleep,
 One plays continually upon a drum.
 They give me bread and water, being a King,
 So that, for want of sleep and sustinance,
 My mind's distemper'd and my body's numb'd,
 And whether I have limbs or no, I know not.
 O, would my blood dropp'd out from every vein,
 As doth this water from my tatter'd robes!

King James I of Scotland and King Edmund Ironside were both killed when sitting in their privies, and Richard III – according to Sir John Harington in his *Metamorphosis of Ajax* – was 'sitting on the draught when he devised with Tirril how to have his nephews privily murdered'.

Harington wrote his *Metamorphosis* in 1596, entreating that it should not be considered 'a noysome and unsavory discourse'. In

SIR JOHN HARINGTON 1561–1612

fact it was an entertaining, satirical diatribe against the filthy habits of his fellow men – habits that he felt he could break with his invention of the water-closet:

When companies of men began first to increase and to make of families towns, and of towns cities; they quickly found not onley offence, but infection, to grow out of great concourse of people, if speciall care were not had to avoyd it. And because they could not remove houses, as they do tents, from place to place, they were driven to find the best meanes that their wits did then serve them, to cover rather than to avoyd these annoiances: either by digging pits in the earth or placing the common houses over rivers.

Harington was dismayed by the dearth of developments with the privy, and was particularly scathing about the fashion of fancying-up the close-stool pan with 'apparell, first disguised to hide nakedness, then applied comliness, and lastly abused for pride: so may I say of these homely places; that first they were provided for bare necessitie . . . then they came to be matters of some more cost . . . for I have seene them in cases of fugerd satin and velvet . . . but for sweetness or cleanlinesse, I never knew yet any of them guilty of it.' He wrote that he had devised a water-closet:

. . . if you have so easie, so cheape, and so infallible a way of avoyding such annoiances in great houses . . . with such savours as where many mouths be fed can hardly be avoided . . . let a publicke benefit expell privat bashfulnesse . . . These inconveniences being so great, and the greater because so generall, if there be a way with little cost, with much cleanlinesse, with great facilitie and some pleasure to avoyd them, were it not a sinne to concele it, rather than a shame to utter it? . . . For when I found not only in mine owne poore confused cottage, but even in the godliest and stateliest pallaces in this realme, notwithstanding all our provisions of vaults . . . of paines of poorfolkes of sweeping and scouring, yet still this same whoresome sawcy stinke, though he were commanded on paine of death not to come within the gates, yet would spite your noses, even when we would have gladliest have spared his company . . . I began to conceive such a malice against all the race of him, that I vowd to be at deadly fewd with them, till I had bought some of the chiefest of them to utter confusion . . . And conferring some principles of philosophy I had read, and some conveyances of architecture I had seene with some other devises of others I have heard . . . I found out at last this way . . . and a marvellous easie and cheape way it is.

It is almost inconceivable to think that Harington had masterminded a proper flushing mechanism – and that no notice whatsoever was taken of it. He laid out 'Enstructions' for his device: 'In the privy that annoyes you, first cause a cisterne . . . to be placed either behind the seat or any place from whence the water may by a small pype of leade . . . be conveyed under the seat to the hinder part thereof . . . to which pype you must have a cocke or washer to yeeld water with some pretie strength . . . Next make a vessell of an ovall forme . . . place this verie close to your seate, like the pot of a close stool . . .' It goes on to describe 'washer stopples' as well as 'a stemme of yron as bigge as a curtain rod'; and recommends 'that children and busy folk disorder it not, or open the sleuce, you should have a little button or shell to binde it doon with a vice pinne.'

The first water-closet had been invented, ahead of its time. As well as deeming it indispensable to comfort, Sir John Harington had realized, 250 years before it was established, that there was a link between disease and poor sanitation – with the 'moiste vapours [that] are apt to spread abroad, and hang like dew about everything'.

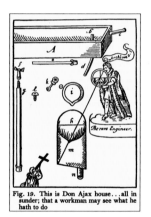

Fig. 19. This is Don Ajax house...all in sunder; that a workman may see what he hath to do

Fig. 20. ...the same, all put together, that the workman may see if it be well

THE FIRST WATER CLOSET HAD BEEN INVENTED, AHEAD OF ITS TIME.

Harington's godmother, Queen Elizabeth I – who had a bath once a month 'whether she needed it or no' – kept a copy of *The Metamorphosis of Ajax* chained to the wall beside the water-closet he had made for her at Richmond Palace. Other than this apparatus and Harington's own, no other – as far as it is known – was ever made, and the Queen greatly preferred her close-stool, which would have been discreetly disguised, as well as comfortingly clad, in rich materials.

Exquisite little velvet chests in appearance, close-stools were far from exquisite in the way they worked, with their softly padded seats bulging around the pot. They were enjoyed by James I, Charles I, Charles II and James II, and at least two still survive today, one at Knole in Kent and the other at Hampton Court. Both are covered in red velvet and have indentical locks, and both are studded with gilt nails galore. One of the first royal close-stools was recorded in 1501; it belonged to James IV of Scotland and, rather curiously, was covered in yards of white cloth. James V of Scotland ordered a green damask 'pavillion' to encircle his 'stool of ease' in 1538. One hopes this means – as it is so pleasing to think of – that this was an ornate little tent. Fifteen and a half ells of cloth (there are 40 inches to an ell) were used to make it up, at the enormous cost of £58 2s 6d (*The Smallest Room*).

Henry VIII had a magnificent close-stool, with the important and interesting addition of 'sesstornes'. It was covered with black velvet, studded with some 2,000 gilt nails and decorated with ribbons and fringes. The seat and the 'elbows' – which addition makes it sound like the finest throne – were covered with white 'fuschan', and stuffed with down. Two leather cases were made, one for the pot and the sesstornes, the other for the chest itself, which suggests that the King took it with him on his travels (*Connoisseur*). In the inventory of Kenilworth, the Earl of Leicester was shown to own twenty-eight close-stools, sixteen of which were of black velvet or satin – both quilted and plain – garnished with either lace or fringes of silver or gold.

Close-stools were certainly more comfortable than their predecessors, the garderobes, but were even less clean, and Charles II and his court seem to have thought it unnecessary to bother with them at all. There is a repellent account by the Oxford antiquary Anthony à Wood of the habits of the King and his 'retinew' while they had been living in Oxford to avoid the plague in 1666: 'To give a further character of the court, they, though they were neat and gay in their apparell, yet they were very nasty and beastly, leaving at their departure their excrements in every corner, in chimneys, studies, colehouses, cellars. Rude, rough, whoremongers, vaine, empty, carelesse' (*The Life and Times of Anthony à Wood, 1632–1695*).

Buckets and pots, sometimes with the added comforts of a cushion or a green baize seat, were the alternatives to which the court had failed to resort. They were still being used, with the same old repulsive consequence of being emptied out on to the streets. One poor fellow, written of in the 1600s, was 'The Foole of Lincoln', who, after his wife had 'so reviled him with tongue nettle as the whole street rung again for the weariness thereof', went out and 'sate him downe quietly upon a blocke before his owne doore' when down came the contents of a 'pissepot' upon his head. ' "Now, surely," quoth he, "I thought at last that after so great a thunder, we should have some raine" ' (*Scatalogical Rites*).

Hobbs, the Tanner of Tamworth in Heywood's play *King Edward IV* (written in 1600), offers the King a chamber-pot of 'fair horn, a badge of our occupation; for we buy no bending pewter nor breaking earth'.

Baths were still being taken in wooden tubs, with scented soaps and herb-filled waters. As soap was taxed and therefore

BATHS WERE STILL BEING TAKEN IN WOODEN TUBS, WITH SCENTED SOAPS AND HERB-FILLED WATERS.

extremely expensive to buy, most people would make their own at home, mixing the fat with rosewater and herbs. To restore the odour of musk (and also the colour of coral) it was advised that you suspended it in the privy for a time (*Medicus Microcosmus, 1600*).

Public bath houses, or 'stews', had been reintroduced into England by the Crusaders, and they flourished until the time of Henry VIII. Then the combination of licentiousness and festering germs, as well as a shortage of wood to light the fires to heat them, forced the King to close them down. They were to reappear, as entirely new and wonderful luxuries, in the eighteenth century.

By the seventeenth century, some progress was at last being made towards a cleaner life. The streets were still filthy, but organized sanitation was slowly improving and efforts were being made to channel water more efficiently into towns and cities. In 1582 a water-wheel had been built under the arch of London Bridge, which gave the water enough force to rise to the height of St Magnus's Church nearby. With the pressure up, the water was piped to the top of the tower, and from so high a point could supply all levels of London (*Clean and Decent*).

In 1609 a new river was brought into London by a Mr Middleton, a goldsmith, who had founded the New River Water Company. He undertook the enormous task of bringing water some 38 miles from the Chadwell and Anwell Springs to reservoirs in Islington, a geographical high point of London. James I granted £8,600 from the Treasury when the finances collapsed halfway through the operation, and Middleton was knighted when water triumphantly poured into London in 1613. In 1690 William Paterson, the founder of the Bank of England, brought water from the springs at Kenwood – or Caen Wood, as it was then called – near the source of the River Fleet, up to the ponds and reservoirs at Highgate. These survive to this day as Highgate Ponds.

With all this increased availability of water, a simple version of the water-closet was at last to appear. Celia Fiennes saw Queen Anne's at Windsor in the early 1700s: 'Within the dressing room is a closet on the one hand, the other side is a closet that leads to a little place with a seat of easement of marble, with sluices of water to wash all down.'

'The Water-Closet' starts to appear in eighteenth-century plans, but only irregularly: in *Convenient and Ornamental Architecture* of 1767 by John Crunden, a water-closet was recommended for only three of the forty-six houses that are shown, beginning with the farmhouse and ascending to the most magnificent villa. The 'closet', also by now 'a general name for any small room' (a *Builders' Guide*, 1770), is suggested for only three houses, as is the privy, and one lot of 'dung holes' is thought suitable for a public inn. The 'Mansion for the Person of Distinction' was to have a water-closet, but there was no convenience at all proposed for an immense country house. A modest villa, it was suggested, should have two privies, but a far grander villa none at all. A 'Town Mansion' should have a water-closet, but there were no such arrangements for 'The Large Town House'. Even 'The Country House of Modern Taste' was to have no fixed sanitation. For 'The Villa of the Palladian Style' it was thought suitable to put the water-closet in the stable wing. It is mystifying that there was such haphazard interest in this invention, now realized and ready to go.

Robert Adam built a water-closet into Shelbourne House in Berkeley Square, and he was responsible for four at Luton Hoo in Bedfordshire. He also designed two at Osterley in Middlesex,

Just in Time

MANY OF THE STREETS, NEVERTHELESS, WENT ON
BEING RECEPTACLES AND RESERVOIRS FOR THE
CONTENTS OF THE CHAMBER-POT
AND THE SLOP PAIL.

both of them in elegant little arched niches. At long last the water-closet was in working existence. However, it was still a primitive apparatus – a marble bowl and a long handle that simply pulled up the plug to release the vessel's contents into the 'D trap', a D-shaped container filled with water that had another pipe leading out from the top of it. This, of course, never emptied properly. These water-closets, though, were the first in a long and unbroken line of inventions, developments and improvements that were to change, slowly but for ever, the grim lavatorial habits that had existed for nearly 2,000 years.

Many of the streets, nevertheless, went on being receptacles and reservoirs for the contents of the chamber-pot and the slop pail. The miseries of walking in London were written of in 1710 by Jonathan Swift in 'A Description of a City Shower':

. . . Returning home at Night, you'll find the Sink
Strike your offended Sense with double Stink.
If you are wise, then go not far to Dine,
You spend in Coach-hire more than save in Wine.

In 1772 London's River Fleet was still being referred to as 'a nauceious and abominable stink of nastiness' – this despite the fact that drains and sewers were being built and that householders were now allowed to connect their private pipes to the main drains, although often at a price.

Smollett, in an account of walking in the streets of Edinburgh in the late 1760s, refers to 'the caution of a man that walks through Edinburgh streets in the morning, who is indeed as careful as he can to watch diligently and spy out the filth in his way, not that he is curious to observe the colour and complexion of the odure, or take its dimensions, much less be paddling in it, or tasting it: but only with a design to come out as cleanly as he may.' There were human lavatories, too, who picked their way through the foul streets of Edinburgh – men carrying pails and wearing immense capes to envelop their customers.

Despite the gradually developing water-closet, the chamberpot continued to be the most popular convenience in the eighteenth century.

Presumptious pisse-pot how didst thou offend?
Compelling females on their hams to bend?
To kings and Queens we humbly bend the Knee,
But Queens themselves are forced to Stoop to thee.

Scatalogical Rites

By now, however, pots were being elegantly disguised inside pieces of furniture – usually a chair or a cabinet, or otherwise a chest of drawers. Chippendale, Hepplewhite and Sheraton all designed bedroom cupboards and commodes, as well as bedside and shaving tables, with ingeniously concealed pots smoothly integrated into their design. Such bowls in the bedroom, or in dank and dark closets, were despised by Swift, who wrote very clearly on how they should be dealt with, in his 'Directions to Servants' of 1745: 'I am very much offended with those Ladies, who are so proud and lazy, that they will not be at pains of stepping into the garden to pluck a rose, but keep an odious implement, sometimes in the bedchamber itself, or at least in a dark closet adjoining, which they make use of to ease their worst necessities; and you are the usual carriers away of the pan which, maketh not only the chamber, but even their clothes offensive, to all who come near. Now, to cure them of this odious practice, let me advise you, on whom this office lieth, to convey away this utensil, that you will do it openly, down the great stairs, and in the presence of the footmen; and, if anybody knocketh, to open the street door, while you have the vessel in your hands: this, if anything can, will make your lady take the pains of evacuating her person in the proper place, rather than expose her filthiness to all the men servants of the house'. Swift also

despaired of those who disguised their chamber pots:

'Ye great ones, why will ye disdain,
To pay your tributes on the plain?
Why will you place in lazy pride?
When from the homeliest earthenware
Are sent offerings more sincere
Than were the haughty Duchess locks
Her silver vase in cedar box.'

PanEgyric on the Dean

He had built two water-closets for himself as early as 1729, which he dramatized in the *Panegyric on the Dean* as being his only achievement of lasting value. The poem is as if written from Lady Acheson, a grateful friend, in whose house they were built:

Two temples of Magnifick Size,
Attract the curious Trav'llers Eyes,
That might be envy'd by the Greeks
Rais'd up by you in twenty weeks:
Here, gentle Godess Cloacine
Receives all Off'rings at her Shrine,
In Sep'rate cells the He's and She's
Here pay their vows with bended knees:
(For, 'tis prophane when Sexes mingle;
And ev'ry Nymph must enter single;
And when she feels an inward Motion,
Comes fill'd with Rev'rence and Devotion)
The bashful Maid, to hide her Blush;
Shall creep no more behind a Bush;
Here unobserv'd, she boldly goes,
As who would say, to Pluck a Rose.

A rose could literally be plucked, as many privies were still being built outside in the eighteenth century. They were simple structures for the poor, elegant little temples for the prosperous – enhancing the gardens in which they stood. A 2-seater and a '2½-seater' – with a smaller seat for a child – both survive in Suffolk, and a '1½-seater' – for the relief of mother and child – is still built into the churchyard of St Mary's, at Selling in Kent. In the garden of Chilthorne Dormer Manor in Somerset there is a little stone building sheltering six seats, with four holes for the grown-ups and two for the children.

By the mid-1700s water-closets had undoubtedly sneaked inside the grander house, although, as we have seen, they were few and far between. Poor George III died in such a room in 1760. Horace Walpole wrote of it to George Montagu: 'He went to bed last night, rose at six this morning as usual, looked, I suppose as if his money was in his purse, and called for his chocolate. A little after seven he went into the water closet, the German valet de chambre heard a noise, listened, heard something like a groan, ran in, and found the hero of Oudenarde and Dettingen on the floor. With a gash on his right temple . . . . he tried to speak, could not and expired'. (*The Smallest Room*)

Baths, too, were an important development of the eighteenth century, with baths and bath houses (little ornate buildings with cosy fireplaced rooms) that were built in the grounds of grand houses. These depended on either rivers or springs for their water source, or upon the tide, as at Antony in Cornwall, where the small blind arched classical building of 1770 has a plunge pool that is open to the skies, but three-quarters roofed over, in the style of a Roman atrium. In the early 1770s Lord Clive of India had an icy plunge pool – a deep marble bath, 12 feet long by 5½ feet wide – built in a vaulted room in the basement of Claremont in Surrey.

Celia Fiennes had found a far more luxurious arrangement at Chatsworth as early as 1694: 'A batheing roome, ye walls all with blew and white marble - the pavement mix'd, one stone white, another black, another ye red veined marble. The bath is one entire marble all white finely veined with blew and is made smooth, but had it been as finely polish'd as some it would have been the finest marble that could be seen. It was as deep as ones middle on the outside, and you went down steps in ye bath big enough for two people. At ye upper end there are two cocks to let in one hot, ye other cold water to attemper it as persons please - the windows are all of private glass'. She also went to the baths at Buxton nearby, where she found the water 'not so warm as milk from the cow'.

Despite all these developments, the eighteenth century remained on the whole a dirty age. There were still very few private baths, although people had started to wash in little tin tubs in front of the bedroom fire. With too few water-closets, the privy and the pot prevailed. The gentlemen had even managed to get the chamber-pot into the dining room for their instant relief, as well as into the billiard room and smoking room, where it lurked behind shutters or was disguised as pieces of furniture. Horace Walpole wrote very funnily of Lord Hervey, the Lord Privy Seal, seeking relief behind a curtain in a room where the ladies were playing cards. He emerged, and 'being extremely absent and deep in politics, he produced himself in a situation extremely diverting to the women. Imagine his delicacy and the extreme passion he was in at their laughing.'

The emptying of all these receptacles was a daily and severe embarrassment with scenes from Swift's 'Directions to Servants' being a miserable reality. In the towns especially, all the pots and buckets had to be paraded through the house, to be collected by the 'night soil men'. This trade was as inelegant as it was elegantly advertised, with cards emblazoned with exquisitely flamboyant script and romantically indeterminate nocturnal scenes, all surrounded by swirls and swags entwined with roses.

Yet, in the 1770s an invention was patented that, by its example, was by and large to put an end to the repellent and unhealthy prevalence of poor sanitation. In 1775 Alexander Cummings, a maker of watches and clocks, took out the first patent for a water-closet: 'a water-closet upon a new construction', with the important feature of the 'S trap', the like of which had never been seen before. This was to be improved two years later by Thomas Prosser, who flattered himself that the 'different noblemen and gentlemen in the three Kingdoms, having used [these water-closets] with satisfaction, will be the means of promoting them' (Patents description). In 1778 this invention was again altered – and this time virtually perfected – by Joseph Bramah, and his version was to remain the most satisfactory water-closet for the next 98 years.

Both Cummings and Bramah were distinguished in their respective fields: Cummings as an horologist and adviser to the Board of Longitude, as well as the inventor of various clocks;

Bramah as the inventor of hydraulic power, no less, as well as the hydraulic press. Bramah also patented rotary engines, fire engines, steam engines and boilers, and perfected printing machines, pens and, of course, his famous Bramah locks, the descendants of which are still in use today. Having started life as a joiner and cabinetmaker, Bramah came upon Cummings's wooden water-closet enclosures inthe late 1770s and, after applying his versatile mind, declared the water-closet's mechanism to be a disaster. The valve – in the centre of the bowl – slid aside and therefore could never be properly cleaned, as it was withdrawn with the flush. This, as well as the hazard of its freezing into immobility – many water-closets were still being built outside – inspired him to invent the flap valve, which was so entirely successful that there are thousands of later copies still being used today. By 1797 Bramah was claiming to have made and sold 6,000 of his 'Valve' water-closets, and the quality of his workmanship was recognized as being so high that the expression 'a Bramah' came to mean anything of first-rate quality.

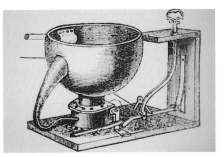

BRAMAH'S 'VALVE' CLOSET

The next water-closet to be invented was the 'Pan', a repulsive and unsatisfactory arrangement that was to be alternately praised and abhorred by all the sanitarians of the day. These men were a new breed who were to meet the needs of the nineteenth century and to improve our everyday lives enormously by their efforts of invention and advice. S. Stephens Hellyer was one such figure, who left excellent, colourful and funny descriptions of the sanitary dilemmas of the day. He not only developed water-closets, he also gave lectures, and wrote several books and pamphlets galore, as well as running the successful firm of Dent and Hellyer. Typical of his prose is this description of how a water-closet could have been nosed out before the new sanitary age had dawned: 'Under the old system, these sanitary conveniences generally advertised themselves, especially in hotels and places of that kind, and all that one had to do in such buildings was to follow the scent like a hound after a fox, by the dictates of an organ which is very useful, but which one does not wish to abuse in

such a way, for, to say the least, it is an offensive way to follow up a thing'. Bathing, he said, with baths and lavatories (as washbasins were then called) should be a necessity rather than a luxury: 'Fitted up with hot and cold services, (they) would, I suppose, be considered a luxury. Well so is a bed, but few John Bulls would care to sleep without one'.

Hellyer was a particularly vicious critic of the 'Pan' closet. It was a foul contraption, with too large a receptacle, into which the hinged pan, attached to the bottom of the bowl, was tipped. The 'puffs of nasty smells', wrote Hellyer, 'which such apparatus send up . . . are enough to make one wish for the old fashioned privy again.'

Through all these vicissitudes, Bramah's 'Valve' was consistently in the background, setting a shining example but never a sales sensation because it was so expensive to produce, although there were several cheap imitators. It was not recorded when the Pan was invented, but it is known to have been 'improved' in 1796 by William Law. It was foul, but what was to follow was just as bad. This was the 'Hopper', a simple cone-shaped vessel – of iron or earthenware – that went straight down to the trap. The surface area was too

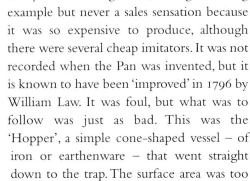

large for the trickling spiral of water to deal with, and it was never properly cleaned. 'There is left a fixture,' wrote Hellyer, that hero of sanitary prose, 'like the basin itself, which the outgoing tenant is generous enough to leave behind him for the incoming tenant to see, and have the benefit of, without anything to pay.' Hellyer suggested that Hoppers could be put to a more useful purpose: 'Instead of destroying the thousands already made, they might be used by market gardeners for protecting rhubarb from frost' (*Principles and Practices of Plumbing*).

There were many sanitary heroes in the nineteenth century: inventors and engineers as well as producers and manufacturers, not to mention all the politicians who had to fight through the clogged inertia of sanitary reform. It was a subject, moreover, that could not be discussed openly in polite society. Mr S. R. Bostel, the grandson of Daniel Thomas Bostel, inventor of the

'Wash-out' water-closet, could remember his grandfather telling him of the great embarrassment suffered by Victorian ladies when faced with his water-closets at exhibitions. 'They would blush and turn away': for them it was like looking at a glistening naked behind, with all its unmentionable associations.

Edwin Chadwick, one of the most prominent reformers of the day, ceaselessly campaigned for a better state of affairs. In 1842 he wrote a *General Report on the Sanitary Conditions of the Labouring Classes of Great Britain*, revealing the atrocious conditions in which families were living all over Britain, and he was in constant battle with Parliament and various authorities. He believed that the dreaded seed of cholera, which struck so viciously in 1831–2 and 1848–9, was nurtured by the filth that lay in mounds at street corners, as well as bubbling through the floors or overflowing up into the outside privies of some 250,000 houses in London alone. In November 1853 *The Builder* reported the case of 'a wealthy man called a contractor' of Bethnal Green, who had in his yard 'a high mound of dirt, filth and stench-matter, as high as the first floor and for some hundreds of yards!' Chadwick was sure that such dirt was the cause of the dread disease, and that a good drainage system, with unadhering stoneware pipes, would kill 'King Cholera' for good. His court still reigned supreme:

'What is my court? These cellars piled
With filth of many a year–
These rooms with rotting damps defiled
These allies where the sun ne'er smiled,
      Darkling and drear!
. . . What are my perfumes? Stink and stench

From slaughter house and sewer
The oozing gas from open trench,
The effluvia of the pools that drench
      Courtyards impure.'

THE BUILDER FEBRUARY 3RD 1855

In 1874 Henry Mayhew gave a startlingly terrible account in *London Characters* of how bad the conditions were. He describes a visit to Jacobs Island in Bermondsey, where the people were as white as vegetables grown in the dark, and the water that ran past their houses was the colour of strong green tea.

As we gazed in horror at this pool, we saw drains and sewers emptying their filthy contents into it, we heard bucket after bucket of filth splash into it, and the limbs of the vagrant boys bathing in it seemed, by pure force of contrast, white as Parian marble. And yet, as we stood gazing in horror at the fluvial sewer, we saw a child lower a tin can, with a rope to fill a large bucket which stood beside her. In each of the rude and rotten balconies, indeed, that hung over the stream, the self-same bucket was to be seen in which the inhabitants were want to put the mucky liquid to stand, so that they might, after it had been left to settle for a day or two, skim the fluid from the solid particles of filth and pollution, which constituted the sediment. In this wretched place we were taken to a house where an infant lay dead of cholera. We asked if they really did drink the water? The answer was they were obliged to drink the ditch, unless they could beg or thieve a pailful of the real Thames.

IN 1874 HENRY MAYHEW GAVE A STARTLINGLY TERRIBLE ACCOUNT IN 'LONDON CHARACTERS' OF HOW BAD THE CONDITIONS WERE.

For centuries the Thames had remained 'the great sewer or

Cloaca maxima for two millions of citizens and for as many more of the population inhabiting the precincts of the paternal stream and its tributaries'. So fulminated *The Builder* in May 1851. Under the heading 'Noisome Reek of the Thames' we are swept into the river's horrors: 'Our tea is infused in it; our viandes cooked; our toddy mixed; our milk watered with it; our beer brewed of it; and every liquid ailment commingles with the filthy exuviae of the foul and ever more foully increasing tide: we lave in it; the body linen of the multitude is steeped therein, and when wrung out the dessicated essences of poison envelop the breathing pores of the wearers. In fact, this corrupt element – which, in the dread epidemic, bore disease in its course, enters into every modification of our sustenance, and we are, despite ourselves, enveloped in its influence, so that the water of life is not the tributary of life but death.' The writer goes on to describe 'other excremental canals which yield their rich tributaries to the tide', and of Fulham Reach, where 'all the horrors of Pandora's box seemed to be amalgated in one steaming vapour.'

Despite the fact that the Chancellor of the Exchequer was reported in *The Builder* of July 1851 to have made the daring statement that 'Sanitary Reform is a Humbug', the reformers struggled on. The Revd Charles Kingsley, author of *The Water Babies*, preached that disease was God's punishment on the populace for leading so repulsive an existence: 'We are calling it a visitation of God . . . yea in the most awful and literal Earnest a house to house visitation. . . . unwholesome habits of living are in the sight of almighty God so terrible and offensive, that he sometimes finds it necessary to visit them with a severity with which he visits hardly any sin; namely, by inflicting capital pun-

FATHER THAMES INTRODUCING HIS OFFSPRING TO THE FAIR CITY OF LONDON.

FOR CENTURIES THE THAMES HAD REMAINED 'THE GREAT SEWER OR CLOACA MAXIMA FOR TWO MILLIONS OF CITIZENS ...

ishment on thousands of his beloved creatures.'

In 1848 a new Public Health Act had been passed, making it obligatory to have a fixed sanitary arrangement of some kind: either an ash pot, or a privy, or a water-closet was to be fitted by every householder. The Metropolitan Commissioners for Sewers, a body founded in 1847, had immediately abolished all cesspits. There had been a 'terrible incident' two years earlier, which was reported in *The Builder* in March 1845:

On the 17th instant, an accident of the most appalling nature occurred in the Female Penitentiary, Exeter. Twenty-one of the inmates retired for a short time to a small room but little frequented, for the purpose of allowing the committee to inspect the apartment they usually occupied, when the floor of the room instantly gave away, and twenty of the unfortunates were immersed in the pestiential contents of an ancient cesspit underneath, the other one supporting herself on a part of the floor still remaining. The cries and appeals of assistance soon brought to their aid the committee, who succeeded in releasing the woman sustaining herself on the broken part of the floor, from her perilous situation and dragging the others from the pit. In five of these however, we regret to say, life was extinct.

In 1858 the Metropolitan Board of Works was founded and, with Sir Joseph Bazalgette at its helm, began the enormous undertaking of laying a sewerage system beneath London. More than 1,000 miles of sewers were to be built; three main routes to the south of the Thames and three to its north. They converged on the East End of London, meeting at Crossness Pumping Station in the south, and at the splendid Byzantine Abbey Mills Pumping Station at Bow in the north. The grand scheme was finished by 1865. London was at last becoming a healthier place to inhabit.

It was George Jennings who first applied the new sanitary technology to public conveniences. He had introduced his ideas at the Great Exhibition of 1851, with his 'Monkey closets' (forerunners of the wash-outs) in the 'Retiring Rooms' in the Crystal Palace, both at Hyde Park and later at Sydenham. He had triumphed against the strongest objections, having been told that visitors were not coming to the Exhibition 'merely to wash'. Jennings was later to be awarded a gold medal for this pioneering and much-needed work, and by the 1890s – progress was slow – he had enriched public thoroughfares all over the British Isles. In his firm's catalogue of 1895 thirty-six towns and 'many others' are listed as boasting public conveniences, from Dundee to Plymouth and from Cardiff to Norwich. The streets of Paris, Berlin and Florence were enhanced by Jennings's public urinals, as were those of Madrid and 'Frankfurt-on-Main', Soulina, Hong Kong and Sydney. He had also supplied water-closets to thirty railway companies in Britain and one in America, as well as others in Buenos Aires, Cape Town and Mexico.

His ideas for the public conveniences were revolutionary; all were built of slate, and many were constructed underground, gracefully marking their whereabouts with cast-iron arches and railings, or sometimes pergolas. Those built above ground were distinctive little buildings in their own right, with finials and pillars or decorative panels, and all were lit by elegant lamps. He also devised a central pillar with urinals around it, which, as well as looking like an architectural flower with its petals open, was economical with both space and water.

All the sanitary manufacturers were to follow suit, although ceramic urinals were to become more popular than slate ones. It took a long time, though, to persuade the people of the advantages of public conveniences. As late as 1882 *The Metropolitan* felt the need to make a plea on their behalf, while at the same time having to tread delicately in leading its readers into this difficult subject: 'While matters are always to be found of which it is not usual to speak in aesthetic society, there are some to which attention must occasionally be drawn, even at the risk of offending our prude propriety. Our readers however, presumably have to deal occasionally with things that do not strictly fall under the "Lily and Dado" category, and they will readily agree with us that the lieux d'aisance, conveniences or urinals, of the Metropolis are of the greatest public utility. London, however, is by no means over supplied with them, and a stranger may walk miles along leading thoroughfares without meeting with one . . . The structure may not . . . call forth rapture upon prolonged contemplation – none of these things do, but if there be an objection, it is one sanctioned by necessity and which has been reduced to a minimum by Mr Jennings, who appears to have a name extending "from Zembla's shores unto far Peru."'

Jennings's catalogues have delightful drawings of the urinals in the townscape, usually featuring a shadowy but elegantly clad figure – featuring just a hint of him adjusting his trousers – emerging from the tiny building. They are beautiful little temples of convenience enhancing wherever they stand, and stirring lines were written to commemorate the opening of the first of such delights:

> I' front the Royal Exchange and Underground,
> Down Gleaming walls of porc'lain flows the sluice
> That out of sight decants the Kidney Juice,
> Thus pleasuring those Gents for miles around,
> Who, crying for relief, once piped the sound,
> Of wind in alley-ways. All hail this news!
> And let the joyous shuffling queues
> For Gentlemanly Jennings' most well found
> Construction, wherein a penny ope's the gate
> To Heav'n's mercy and Sanitary waves
> Received the Gush with seemingly, cool obedience,
> Enthroning Queen Hygeia in blessed state
> On Crapper's Rocket: with rapturous ease men's cares
> Shall flow away when seated at convenience!

THE GOOD LOO GUIDE

Developments within the home were seldom of an order to call forth such a eulogy. The efficiently flushing water-closet might well be found in one house, while next door you would have to trudge off to the privy at the bottom of the garden. Many wash-basins had become objects of grace by the 1880s, with ornate cast-iron stands supporting fancifully framed mirrors that soared above the basins like great espaliered trees, splayed flat against the wall. Immense canopied baths – little wooden houses of carved or mirrored panels – were starting to appear in affluent surroundings, although in the vast grandeur of Belvoir Castle in Leicestershire all washing was still done with jugs and basins at the wash-stand, and with tin hip-baths in front of the fire. Baths were not installed at Belvoir until 1912, and Lady Diana Cooper, who lived there as a child, described the 'water-men' in *The Rainbow Comes and Goes*:

The unearthly watergiants, who were always moving about the house keeping the jugs, baths and kettles full . . . The water-men are difficult to believe today. They seemed to me to belong to another day, they were the biggest people I have even seen . . . They wore brown clothes, no collars and thick green baize aprons from chin to knee. On their shoulders they carried a wooden yoke from which hung two gigantic cans of water. They moved on a perpetual round. Above the ground floor there was not a drop of hot water and not one bath, so their job was to keep all jugs, cans and kettles full in the bedrooms, and morning or evening to bring the hot water for the hip baths. We were always a little frightened of the water-men. They seemed of another element and never spoke but one word, 'water-man', to account for themselves.

The gas-heated bath of 1871 – with naked flames burning

... PUBLIC CONVENIENCES WERE REVOLUTIONARY; ALL WERE BUILT OF SLATE, AND MANY WERE CONSTRUCTED UNDER-GROUND, GRACEFULLY MARKING THEIR WHEREABOUTS WITH CAST IRON ARCHES AND RAILINGS ...

beneath an ornate metal tub – was a terrifying alternative of the times. One, called the 'General Gordon Gas Bath', had a little pinnacled canister connected to it, in which you warmed your towels. There was a hinged bunsen burner that swung under the bath once its flames were lit. But as there was no flue the combination of the scorching hot metal and the gas vapours must have made taking such a bath a misery.

Water-closet inventions were appearing all the time. In 1844 a Mr Wiss of Charing Cross patented 'The Portable Closet', a wooden chest with a little cistern of water and a bowl all neatly enclosed together. This was taken up and sold by all the suppliers until the end of the nineteenth century. The Revd Moule's 'Earth Closet' of 1860 was another contraption, and a curious one at that for so late a date, but nevertheless it proved to be a great success. It was based on the optimistic belief that dry and sifted earth or sand were excellent deodorizers, and that they stayed ever sweet, however many times they were used. A cone behind the seat was filled with earth or sand which could then be released down a chute, in a measured amount, whenever the handle was pulled.

The use of 'the excretia of the "paragon of animals" . . . on scientific, moral, sanitary, and economic grounds' was advocated in 1897 by Dr Vivian Poore in his book *The Dwelling House*. He had 'rather handsomely manured his garden with human excretia' and recommended the three most suitable systems to that end: the 'Indoor Earth Closet'; the 'dry catch', which entailed simply a slope of concrete beneath the seat of an outside privy; and the 'pail method', which was the least satisfactory. Today, the ecological lavatory is enjoying an ever more respectable following. It consists of a box, creating optimum conditions for compost and more often than not has specific plants growing nearby, providing leaves to use instead of paper. Many thousands are now being built in Australia.

Water-closets, as well as baths and 'lavatories' (as wash-basins were known), at last primed for popular appeal, were to become objects of rare beauty during the exhilarating years of their heyday between 1875 and 1900. The water-closet in particular – that messiah of sanitary innovation – was given the most magnificent attention of all. It was wrought into a medley of shapes, with innumerable colours and colour combinations, and as many designs and decorations as was possible to conceive, crammed on to those single ceramic or cast-iron pedestals and bowls. You could sit on a wave spraying from a dolphin's mouth or relish relief on a key-patterned bowl perched on the back of a lion! All this was, no doubt, partly to disguise the water-closet's unmentionable function, but in the main it was a loud decorative huzza for this saviour from sanitary squalor.

It seems to have been the Prince of Wales's near death from typhoid in 1871 (Prince Albert, his father, had died of the disease ten years earlier) that jolted public consciousness on sanitary reform. The Prince himself had been so stirred by the possible consequences of bad drains that he declared that he should like to have been a plumber if he were not a prince. Certainly, it was from this point that domestic sanitary innovation flourished. 'How many deaths', wrote our old friend Hellyer, 'have been caused by a polluted water tank, a brick cesspool, a foul drain, diseased water-closet trap, a bottled up soil pipe, a sink 'bell' trap? Is not the very name ominous and ought it not rather be called

'a death bell trap'? At any rate it sets the death bell ringing occasionally. All England was alarmed some time ago, when it heard one of these gates to death rattling upon its hinges and threatening at every moment to fly open – under the influences of one of the evils innumerated above – for a Royal Prince to pass through. Another puff of bad air, and who knows how wide the gate would have opened.'

It was Hellyer who was to improve Bramah's 'Valve' closet – some 100 years after it had been invented – with his 'Optimus'. He produced sixteen of them with such variations as 'bellows lined with calf skin'. 'They have several imitators,' wrote Hellyer, '. . . but still remain – OPTIMUS.' They were produced until the outbreak of the Second World War and could boast of a most distinguished patronage: Queen Victoria installed them at Buckingham Palace and Windsor as well as Holyrood House and Osborne, and Edward VII was to have them at Balmoral and Abergeldie Castle. They were also enjoyed by the Duke of Wellington, George V, the Tsar of Russia and the King of Siam. Every one of the original water-closets in the Houses of Parliament was an Optimus, as well as those in the Royal Courts of Justice and the War Office. Among the forty-seven patents that he took out, Hellyer always claimed to have invented the 'Wash-out' – after having seen Jennings's Monkey Closets based roughly on the same principles. None of Hellyer's models seems to have survived, though, and no more was written of them elsewhere.

Daniel Thomas Bostel of Brighton, however, was exhibiting his 'Excelsior' wash-out as early as 1875, and could justifiably

ALFRED GOSLETT & CO., GLASS AND COLOUR MERCHANTS, 103

GAS BATHS.

THE "MARVEL" GAS BATH, Right Hand.
Fitted with Plug and Washer and Union. Japanned Oak outside, White Marble inside.
No. 1674. 5 feet .. £4 19 6 | No. 1675. 5 feet 6 inches .. £5 14 6

THE "GENERAL GORDON" GAS BATH.
This is a well made Bath, the sides being of best sheet Iron, tinned and japanned white inside, dark oak outside. The bottom being copper is a good conductor of heat, and is so little affected by the gas as to be much more durable than iron would be.
It is fitted with Waste Plug and loose Bent Union for attaching pipe, movable Towel Warmer, and powerful Atmospheric Burner. Price complete, including Gas Tap—
No. 1676. 5 feet .. £5 10 0 | No. 1677. 5 feet 6 inches .. £6 0 0

THE "PRINCE OF WALES" GAS BATH,
With Semi-Roman top, and Copper Boiler, fitted underneath with patent safety Atmospheric Burner, and Linen Airer attached to bath.
Japanned Green Marble outside.
Do. White do. inside.
No. Length. Top. Foot. £ s. d.
1678. 5 ft. 6 in. 24 in. 20 in. 12 18 0

26, SOHO SQUARE, AND GEORGE YARD, CHARING CROSS ROAD, LONDON, W.

THE GAS-HEATED BATH OF 1871 – WITH NAKED FLAMES
BURNING BENEATH AN ORNATE METAL TUB –
WAS A TERRIFYING ALTERNATIVE OF THE TIMES.

claim to have been its inventor. His grandson had one of the original pedestals until he died in 1989. His great-grandson still runs the firm of Bostel Brothers, which has been going strong since 1789.

The next wash-out to appear, this time with its works hidden away by a wooden enclosure, was the 'National' of 1881 by Thomas Twyford. This had such a commercial success that by 1889 his firm could claim that 100,000 were in use. It was followed in 1883 by an all-ceramic pedestal version, the 'Unitas', hailed as a triumph both technologically and decoratively (it had an oak tree in raised relief growing up its bowl). Thousands of these WCs were exported abroad, and in Russia the word 'unitas' became part of the language, meaning excellence.

'Unlike ordinary W.C. basins, it is not enclosed with woodwork,' trumpeted Twyford, 'but is fully exposed, so that no filth nor anything causing offensive smells can accumulate or escape detection.' There were several tests at this time to track down where smells came from: the 'Peppermint Test', as well as the 'Gas' or 'Ether', or the 'Smoke' and 'Ammonia' tests. The escaping stinks were nosed out with canisters — filled with the oils, gases or smoke — on long handles that were thrust between the two traps to pinpoint escaping fumes.

The first Thomas Twyford — direct descendant of Joshua Twyford, manufacturer of teapots in 1700 — made sanitary ware in Staffordshire until his death in 1872, when his son Thomas William Twyford took over the firm. The firm's designs became dazzling, forging into this new field with a vim and a verve that astound. Its catalogues alone are like illuminated manuscripts, with page after page of exquisite drawings

ITS CATALOGUES ALONE ARE LIKE ILLUMINATED MANU-
SCRIPTS, WITH PAGE AFTER PAGE OF EXQUISITE DRAWINGS
AND A TOUCH OF GUILDING HERE AND THERE AMONG
THE SUBTLE COLOURS.

and a touch of gilding here and there among the subtle colours.

Edward Johns was another sanitary innovator, who made one of the finest WCs of all — 'The Lion Pedestal' of 1896, with a handsome key-patterned bowl on its back. In 1900 his pottery was sold to Alfred Henry Corm, who thought up the sales wheeze of emblazoning 'Not by Royal Appointment to the King' on to his dolphin-shaped WCs.

The pottery became known as Armitage in 1960, and in 1969 it merged with the grand old firm Shanks of Barrhead. John Shanks was another of the great sanitarians, who started business as a plumber and gas fitter in Paisley. In 1864 he patented his first WC — the 'No. 4', an improved version of the old plunger closet, with a valve and ballcock — and by 1866 he had built a vast new brass foundry called The Tubal Works. At the first testing of his No. 4 Shanks had seized the cap from a workman's head and flung it into the bowl, gleefully shouting 'It works!' as his poor employee's cap was flushed away. By 1895, when he died, his firm had become a worldwide concern, making beautiful baths and water-closets as well as 'lavatories' of the finest quality of workmanship, inventiveness and imagination. Its catalogues, too, make you linger long and lovingly over the pages.

Doulton produced these elegant books as well, one of whose pages was devoted to green 'marble' urinal stalls supported by giant dolphins. The firm had been founded by John Doulton in 1851, but it was his son Henry who was to contribute so enormously to the world of sanitation. He produced the first stoneware piping for London's drainage system, after he and Edwin Chadwick had waged a long and successful war against porous and unhealthy earthenware pipes. A delightful story survives of their friendship. Chadwick, in a great state of excitement, had come to see Doulton: 'I have made a great discovery, my dear friend . . . A pig that is washed puts on one-fourth more flesh with the same amount of food as a dirty pig. We must have automatic, and if needful, compulsory washing for the English lower classes. I have devised an apparatus by which one man may be completely washed, even against his will, in five minutes, a woman in four and a child in three. It can be done, with the help of an apparatus you will manufacture for me, at the rate of ten a penny. England clean and fat, my friend, at the laughably trifling cost of one penny for ten adults!'

Sir Henry Doulton died in 1897, ten years after receiving his knighthood. His firm, though, was to go on from strength to strength. The Doulton catalogue of 1916 could boast of a most exotic and distinguished patronage: 'The Residence of H.E. the Viceroy of India' and 'The Viceregal Lodge, Delhi', as well as the Residences of the Governors of Madras, Bombay, the Punjab, Bihar, Orissa and Bengal. Doulton also supplied the High Courts in Delhi and several palaces, including 'The Nao Talao Palace in Gwalior' and the 'New Palace at Ujjain, for H.H. the Maharajah Scindia of Gwalior'. Then there was 'The Palace of Rutlam, for H.H. the Rajah of Rutlam', as well as palaces at Dholpur, Benares and Cooch Bihar. Royal Doulton – it was given the Royal Warrant in 1901 – was one of the many firms to ensure that Britain ruled the sanitary waves!

George Jennings, as well as inventing the first public convenience, was one of the greatest of all these innovators. As early as 1847 Prince Albert had presented him with the Medal of the

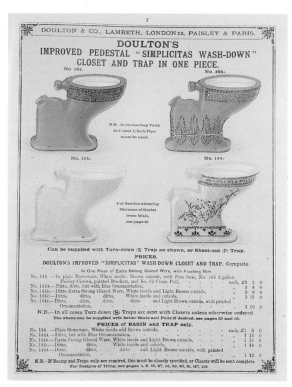

THE DOULTON CATALOGUE OF 1916 COULD BOAST OF A MOST EXOTIC AND DISTINGUISHED PATRONAGE.

Society of Arts. 'The Prince Consort greatly encouraged this indefatigable Engineer,' proclaimed the *Sanitary Record* in 1897. 'In sanitary science he was avant coureur in his day and generation, and was among the first Engineers to practically carry out the theories of the wise men of the time. "Sanitas sanitarium" was Mr Jennings's motto before Disraeli adopted it as his political maxim.' He had implored a shocked city of London to accept his public lavatories free, on the condition that the 'attendants whom he furnished were allowed to make a small charge for the use of the closets and the towels. But the venerable city fathers would not.'

Jennings fitted up the hospitals at Scutari and Varna. At St Paul's Cathedral he supervised the sanitary arrangements at the thanksgiving service in 1872 for the Prince of Wales's recovery from typhoid. He provided the Empress Eugénie of France with a copper bath, and supplied the former Khedive of Egypt with a great multi-showered contraption with a mahogany canopy. For hospitals and asylums he devised a bath with a lever-action supply valve that ensured that the hot water could never be let into the bath unless the cold was first and proportionately turned on. It was Jennings who patented the popular 'tip-up' basin that swivelled round on an axle to be emptied; and, of course, there was his syphonic 'Closet of the Century'. A Mr J. R. Mann had already patented the syphonic system in 1870, but with no great success; Jennings's improvements were to make the syphonic model one of the most prestigious lavatories of its time.

At last the word 'lavatory' can be used in the modern sense – until the turn of the century it always referred to the wash-basin or the laver, the lavabo, the lavatorium etc.

We have George Jennings to thank for producing the direct ancestor of our lavatories today, with his 'Closet of the Century'; and the firm of Humpherson and Co. of Chelsea, for first patenting the 'Wash-Down' WC with their Beaufort of 1884. This firm had been founded in 1876 by Edward Humpherson, who was later joined by his two sons, Frederick and Alfred, after they had served apprenticeships with Thomas Crapper at his Marlborough Works in Chelsea. Humpherson, at its Beaufort Works, produced a number of patented inventions, winning prizes and medals at exhibitions, but its greatest achievement of all was the 'Beaufort'. This was the first patented pedestal wash-down WC, from which so many millions have since been copied, most of them still in hard-working use today.

Thomas Crapper, nearby, was in constant professional rivalry with Humpherson. Crapper's main interest – contrary to the belief throughout the world that he was the be-all and end-all of sanitation – was cisterns and 'water waste preventers'. It seems that he did supply one lavatory, emblazoned with the Prince of Wales's feathers, after having been given the Royal Warrant when he installed the drains and sanitary fittings at Sandringham between 1886 and 1909. The firm of Thomas Crapper and Co. Ltd was still going strong in 1935, but it was finally taken over in 1950 by John Bolding, yet another distinguished firm that had been founded as early as 1822. It was Bolding that supplied the wave-spraying dolphin, one of the most imaginatively designed lavatories of all, and it also sold a bath with four legs resembling dolphins that slithered down from the rim to the floor. Its 'Pillar' pedestal closets – three pillars in a cluster, often made of enamel-covered iron – were very successful.

The Honorary Testimonial and MEDAL OF THE SOCIETY OF ARTS, *and the* EXHIBITION PRIZE MEDAL OF 1851, *were awarded to* GEORGE JENNINGS, *for his various* PATENTED SANITARY INVENTIONS.

JENNINGS' ·
PATENT TILT-UP AND LIPPED LAVATORY,
For Hospitals, Barracks, Asylums, Schools, Stations, Unions, and Buildings of every Class.

IT WAS JENNINGS WHO PATENTED THE POPULAR 'TIP-UP' BASIN THAT SWIVELLED ROUND ON AN AXLE TO BE EMPTIED …

George B. Davis, who patented a ventilating system for the lavatory, as well as urinals and syphonic closets, was yet another figure of importance in this field. It was he who produced the exquisite 'Albania' wash-down, with trumpet-blowing cherubs moulded in relief around the pedestal – described in the *Plumber and Decorator* of 1899 as 'a most handsome lavatory'. Then there was the firm of John Tyler and Sons of London, which was one of the earliest to be founded, in 1777. It was later to produce baths and lavatories of all types, one of the most charming being a high-backed, woodgrain-painted tin bath on wheels, to be pushed with a long handle. It was mainly for hospitals and could be wheeled about without a sound, on its vulcanized india-rubber 'tires'.

Another distinguished firm was Matthew Hall, which supplied a most elegant WC, shaped like a swan with a bowl on its back, on which you sat. The firm fitted up the City Club in London with 'Lavazonic' wash-basins and 'Sylph Syphonic' lavatories. Adamsez was another exceptional establishment, with a reputation for the highest craftsmanship. It produced two very curious contraptions: a remarkably futuristic cistern that was flushed by pressing a button, and a wash-basin that swung out over the lavatory to save on plumbing. Its sanitary ware was reputedly of the highest quality, and it is still remembered with admiration.

Between them – in their happy heyday – all these firms produced hundreds of innovations and variations on the water-closet theme. There were 'Trapless Twin Basin' closets, and 'Ventilating Pan' closets, as well as those with a 'Treadle Action'; and also 'Pneumatic' and 'Pneumatic Combinations'. Then there were the 'Elastic Valve', the 'Flusherette Valve', the 'Valve Hopper' and the 'Pan Valve', not to mention the 'Trapless Valve' and the 'Water Battery', and the closet with the 'Self-Acting Seat'. And there were numerous other frightening-sounding 'Self-Acting Closets'. Today only four WCs are made in the British Isles: the 'Wash-Down', the 'Syphonic', the 'Squatting Bowl' and the 'Wash-Out', which is still produced for suppliers abroad.

The astonishing thing is that all this happened so recently: the lavatory as we know it today was invented little more than 100 years ago. Since the 1880s it has changed neither its workings nor its basic shape. Modernization has simply meant a streamlining of what was a rich, delightful and enjoyable form. The lavatory is an intimate friend to us all, and we should honour it as such. It is undeniable that a glorious throne with a welcoming wooden seat makes us laugh with pleasure. Why then do we minimize its importance, treating is as a mere receptacle, a necessary evil?

But there is a ray of hope. After seventy years of sterile sanitary design the bathroom is at last becoming the revered room that it should be. Architects and designers, realizing the 'sheer, solid joy' that a well-designed lavatory and a capacious bath can give, are once again planning temples in which we can luxuriate.

Such was the state of affairs in 1977, when the Avocado-coloured suite still reigned supreme, but when a realization was at last dawning that design in this field should and could be improved. Eighteen years later these dreams have indeed been realized; adventurous and innovative designers have sent many a lavatory streaking off into the future, and a long overdue appreciation of nineteenth-century sanitary splendours has produced a new flowering of ornate patterns and pans. The designs of baths, as well as basins and bidets, taps and tiles, are now all displaying an attention to detail that would have seemed inconceivable in the fifties, sixties and seventies. As a miniature mirror of taste today, contemporary sanitary design is hearteningly happy.

James Williams, with his firm Sitting Pretty in Fulham, was the first to dip his toe into the warmer waters of sanitary design in 1982. He saved many a valuable piece from inevitable destruction and produced innumerable designs of his own. His 'Priory' is a triumph of restrained relish for the past, along with contemporary streamlined simplicity. John Anderson of Anderson

Ceramics was next in line to beautify the lavatory, with his Imperial Suite of 1981, showing writhing floral forms in deep relief, again designed by James Williams.

Many were to follow suit, and today such firms as Sanitan and Vernon Tutbury, as well as The Imperial Bathroom Company, Caradon Twyford, Heritage Bathrooms and Chatsworth, are all producing quantities of elegant designs. Many firms are supplying the very best of these, such as Sitting Pretty and Original Bathrooms in Kew, which is owned by Geoffrey Pidgeon – the great nephew of the inventor of the wash-down. He also sells 'Old England', a range that has been produced in Italy by Simas, after '. . . a profound study of past bathrooms'. Other adventurous firms abound in Britain today, such as The Water Monopoly in London which rescue's and restores vast quantities of sanitary art from the eighteenth, nineteenth and twentieth centuries.

Sadly, the strong colours that blazed forth from early sanitary ware are no more. Thanks to their deadly content of lead, they are now illegal and all patterns today have to be transfer printed. This obstacle will no doubt soon be overcome but meanwhile white porcelain wins hands down in the 1990s, designed into elegant forms and embellished with moulded decoration. 'The Priory' is a glowing gothic testimony of its possibilities.

For good modern forms, such architects as John Pawson and John Young have produced sheer and simple splendour in the bathroom. For more mass comfort, it is Santric of Swindon which streaks ahead of all the rest with its smooth and sheer stainless steel lavatories and washbasins, designed for the disabled as well as for hospitals and prisons. The prize for the best public Gents and Ladies – indeed it was the best building to rise up in 1993 – goes to Piers Gough's entrancing creation in Westbourne Grove in London. This is a story of local heroism on the grand scale: John Scott of the Pembridge Association, seeing that the public lavatory was to be replaced, offered to have it designed by an architect. The estimate was some £10,000 over what the council could afford and John Scott himself paid the difference. An exquisite little structure emerged, a slender triangle that quite slices itself into your senses. What triumphal hope for our lavatorial future!

THE PLATES

The latrine at Housesteads, (previous page) the great fort on the Roman Wall in Northumberland, where as many as twenty men would sit side by side, relishing, rather than recoiling from, each other's company. A continuous wooden seat was built around and over the deeply dug outer sewer. Beneath the sitters' feet was a smaller watercourse, into which they dipped their monstrously, unappealingly communal sponge sticks – which they used instead of paper. There being neither a well nor springs nearby, rainwater was all they had for flushing. It flowed into both channels from great stone tanks, and then on, through pipes, down the hill, where the manure was collected, and the liquid – literally 'liquid gold' – was poured into pots for the fulling of cloth. It is thought that the stone basins were used for washing hands.

The fifteenth-century lavatorium in Gloucester Cathedral (right) is housed in a miniature version of the great building's fan-vaulted cloisters. Hidden away until you peer round its arches, this splendid, once lead-lined, wash-basin is for-ever splashed with the hues of the stained glass that surrounds it. The drain holes, through which water once slurped into the Fulbrook stream below, still survive.

The steamy splendour of the Great Bath (opposite) at Aquae Sulis (as Bath was called), built by the Romans in the late first century AD but abandoned by them some 200 years later. In the 1870s, after nearly 2,000 years, they were rediscovered, miraculously still intact, and were gradually restored to their former splendour. The lead pipe in the foreground is original, but every-thing above the surviving Roman pillar bases is Victorian.

The late fifteenth-century garderobe at Cleeve Abbey in Somerset, built into the thickness of the wall that supported the Frater pulpit above. Cleeve Abbey was founded in 1142 by the Cistercians, who most wisely built it over the River Washford, which provided an excellent sluicing and draining system as well as refreshment for the living larder of their fish pond.

Part of the twelfth-century laver (opposite), once a richly carved shrine to cleanliness, at Much Wenlock Priory in Shropshire. Originally enclosed by a small circular pillared building, the laver rose up in three tiers around a central column, from which little stone creatures spouted water into the washing-basin below. Christ and the Apostles on the Sea of Galilee are carved rowing around the base, and can still be seen, in clearest detail.

A medieval lavabo carved with a hound chewing a bone in the undercroft of Wells Cathedral in Somerset. Its date is not known; the undercroft was built between 1170 and 1320, and a quantity of the most curious carvings were produced, including fighting dragons as well as snakes and with human male and female heads.

Nine arches for nine latrines (opposite), enhancing Fountains Abbey in Yorkshire. They were built between 1160 and 1180 for the lay brothers, between their dormitory and the infirmary. The Cistercians, who founded the Abbey in 1132, built three immense reredorters at Fountains, all of them arched and picturesquely ornamental, and all of them sensibly sited hard by the River Skell.

The grotesque open-mouthed face, through which all the filth flowed from the garderobes at Beaumaris Castle, on the island of Anglesey in North Wales. The castle was built by Edward I between 1295 and 1330.

The garderobes – corbelled out for cleanliness – on the fourteenth-century tower of Chipchase Castle (opposite) in Northumberland. The only hope of keeping odours at bay was to build long and narrow passages – within the thickness of the walls – to these convenient little chambers. Henry III had to order double doors for his garderobe, when his walls proved not thick enough.

One of the two-seater garderobes at the scrupulously restored Haddon Hall in Derbyshire. Beneath the seat is a drop of some 30 feet, on to a steep slope of rock below, where either the rain or a miserable menial would wash the filth away.

No fewer than three garderobes and two latrines can be seen (opposite) in this view of thirteenth-century Greencastle in County Down, Northern Ireland. Two latrines survive in the ruins of the surrounding walls, with their outlets to the moat cut deeply into the rock. The three projecting garderobes that stand pleasingly clear from the walls – otherwise there would be the vilest staining – date from the sixteenth century, when the tower was rebuilt.

There is a host of seven-teenth-century chamber pots to be found in The Museum of London. They are of earthenware, some dating from as early as 1600; the oldest is on the bottom right, glowing in 'Tudor Greenglaze'. These shapes stayed the same through-out the century; the taller the pots are, the earlier their date.

Three 'Metropolitan' slipware ornamental pots, dating from the period 1630–80, with pipe clay decoration that has turned yellow under the lead glaze. They were mostly made around Harlow, then given the name 'Metropolitan' when they were sent to London.

A six-seater, no less, at Chilthorne Dormer Manor, near Yeovil in Somerset (opposite), dating from the late seventeenth century. The seats – made of soothingly soft oak – march round the panelled walls of the enchanting little stone building, with a stone ball atop its pyramid roof, at the bottom of the garden. There was room for four grown-ups and two children, and it was a daily ritual for them all to gather together. I was told of a similar scene at the Belfast Cattle Market in 1937, when six farmers, all with their trousers down, sat two by two, each man holding the page of a broadsheet newspaper. Four were reading the papers, the other two were having a good chin-wag.

A rich and rare royal close-stool, covered in red velvet and with a stuffed horsehair seat, at Knole, near Sevenoaks in Kent. It was probably used by Charles I and Charles II and was certainly sat on by James II. Similar

sumptuous stools were made for other monarchs. Henry VIII enjoyed such comforts in 1547 with 'The vse of the Kynges Mageste', which was covered with black velvet and 'garnished with ribbon and nails and fringes'. The seat and 'elbows' were covered with white 'fuschan' filled with down, and 2,000 gilt nails were used for 'garnishing'. It was supplied with leather cases, one for the pot, and the 'sesstornes' for the chest itself. Only two of these royal relics are known to survive: this one at Knole, and another, also covered in red velvet, at Hampton Court.

The eighteenth-century Chinese dressing room (opposite) at Saltram in Devon. It was always used for bathing and dressing, as you can see from the original screen, built to provide both privacy and warmth for the bather. The wallpaper, with figures the size of children, dates from the reign of K'ang Hsi (1662 –1722) and was put up when Saltram was finished in the mid 1700s. The cans for hot and cold water, as well as the bath and footbath, are all Victorian.

Wooden two- and three-seaters, curiously alike yet some 20 miles apart from each other in East Anglia. The two-seater is at Thorpe Hall, Horham, near Eye; the three-seater (opposite) at Hempnalls Hall, Cotton, near Stowmarket. Both are of pine and could be of any date from the early eighteenth century to the late nineteenth. The buckets still survive at Thorpe, ever ready to be used and then emptied into the moat below.

The water-closet in the Turret Room at Osterley (opposite) in Middlesex. The house, originally built by Sir Thomas Gresham in the mid-sixteenth century, was transformed two centuries later into a curious classical palace by William Chambers and Robert Adam for Sir Francis Childs, grandson of England's first banker.

There has always been a water-closet in this niche in the Turret Room. It is recorded that a carpenter called Matthew Hillyard installed a 'marble closet pan' with a great plug, as early as 1756. In the 1870s this was miraculously modernized when Dent and Hellyer installed their 'Optimus' Valve closet – a curious choice, as its only advertised advantage was a 'slop top'. 'Splashings', it was claimed, could not 'find their way over the basin and cause unseen nuisance within the enclosure: the unpleasant draught which is frequently experienced when using enclosed closets is also prevented.'

Marie-Antoinette's gilt-enriched Sèvres porcelain travelling pot of 1758, which was most delicately decorated by Binet. It is tiny – only 6 inches in diameter – and would have kept discreet company with the Queen in its little black and gold Chinese box. A poignantly pretty object, it is seen here at Blenheim Palace, on a Louis XV table, beneath portraits (both by Reynolds), of Elizabeth, Countess of Pembroke, and Francis, Marquis of Tavistock.

An Angoulême porcelain pot, with Marie-Antoinette's entwined initials on the outside, and a little peeing child on the inside. It was made by Dihl and Guerhard, of Paris, in the 1780s, with an elegance that belies the vilely inelegant habits of the day. In *Humphrey Clinker*, published in 1771, Tobias Smollett wrote a delectable description of how the Scots employed their chamber-pots:

> And now, dear Mary,, we have got to Haddingborrough, . among the Scots, who are civil enuff for our money . . . But they should not go for to impose upon foreigners; for the bills in their houses say that they have different EASE-MENTS to ley; and behold there is nurro geaks in the whole kingdom, nor anything for pore sarvants, but a barrel with a pair of tongs thrown a-cross; and all the chairs in the family are emptied into this here barrel once a-day; and at ten o'clock at night the whole cargo is flung out of the back windore that looks into some street or lane, and the maids call GARDY LOO to the passengers, which signifies LORD HAVE MERCY UPON YOU! and this is done every night in every house in Haddingborrough; as much as you may guess, Mary Jones,

what a sweet savour comes from such a number of perfuming pans; but they say it is wholesome, and, truly, I believe it is; for being in the vapours, and thinking of Issabel and Mr Clinker, I was going into a fit of astericks when this piss, save your pres-ence, took me by the nose so powerfully, that I sneezed three times, and found myself wonderfully refreshed; and this to be sure is the raisen why there are no fits in Haddingborrough.

This pot is now at Newby Hall in Yorkshire, which has an interior by Robert Adam, with a classical sculpture gallery, as fine as any in Europe. It also has a rare set of Georgian and Victorian chamber-pots, which were collected in the nineteenth century by Robert de Grey Vyner.

The marble bath, 5½ feet deep and 12 feet wide, which was built for Lord Clive of India in the early 1770s at Claremont, near Esher in Surrey. It was not for pure luxury, as Clive had been ordered by his doctors to take cold plunges to treat a nervous complaint. This elegant pool was built, tuning into the tastefulness of Henry Holland's interior for Capability Brown's great classical house. The bath is in the basement, in a little vaulted room that is now a classroom, still with its stucco medallions on the walls. A wrought-iron railing once decorated either end of the plunge pool, which was built at a cost of £310 15s 3d (£310.76).

Claremont has the most romantic of histories. Poor Princess Charlotte – the Prince Regent's daughter, who died in childbirth in 1817 – lived here for her last eighteen months, with her husband Prince Leopold of Saxe-Coburg. Queen Victoria loved the house, and stayed there both as a child and later with Prince Albert. She had a lifelong interest in Claremont and gave it to her son Leopold, Duke of Albany, in 1883. His daughter Princess Alice would always remember with delight the sight of Ruskin prancing around the hall to the sounds of a barrel organ, being played by her father. All these people might well have

used the bath. The exiled King Louis-Philippe and Queen Marie-Amélie of France most certainly did. She lived there with her children and grandchildren for eighteen years, from 1848 to 1866 (the King died in 1850), and the French royal children were brought up with the most rigorous and spartan regime of cold baths, gymnastics, swimming and even lessons in the art of war.

The house is now Claremont School for girls.

'Seats of ease' on board HMS Victory, the great flagship of the fleet at the Battle of Trafalgar. There were only six such 'heads' – the nautical name for such conveniences, as they were built in the bows of the ship – for the five hundred crewmen. All the officers had garderobe-like conveniences, hanging out over the ship's stern from 'quarter galleries' while Lord Nelson enjoyed the charms of 'counterfeit cabinet'. This was an elegant piece of furniture, with a multitude of drawers, each with a lion head handle, and with one that pulled out to reveal a chamber pot. The exact same piece of furniture is to be found at Biddick Hall in County Durham, see page 53.

On Victory, the 'seats of ease' were conveniently close to the sick bay although a good tramp – some halfway across the ship – from their sleeping quarters on the lower gun deck. The ship was scrupulously clean, with the decks being regularly scrubbed with 'Holy Stones' – so called because they were the same size as family Bibles. Of the whole English fleet, Nelson's Mediterranean Fleet had the best sick record, with a good diet, as well as cleanliness, being of prime importance. In 1804 alone the men consumed 21,000 oranges and 81,685 onions. When Victory was built between 1759–1785, over two thousand oak trees were felled to construct her hull. Twenty-seven miles of rigging were strung aloft, and hung with acres of sail. She had cost £63,176, the equivelent of £50,000,000 today.

Three conveniences for the comfort of gentlemen after dinner:

The late eighteenth- to early nineteenth-century pine column commode that was sold

from Raby Castle in 1994 after providing years of cosy classical comfort to the gentlemen after dinner.

The William IV mahogany and ebonized pot cupboard at Raby Castle (left), in County Durham. When photographed it was in the library, as the dining room was being repainted, but normally it would have stood by the sideboard, to be made use of by the gentlemen once the ladies had retired after dinner. The portrait (c.1680) of Lady Mary Sackville is by Lely.

This early nineteenth-century pretending plate bucket (opposite), graciously concealing the chamber-pot – again for relief after dinner – is in the dining room at Seaforde House, in County Down, Northern Ireland. The portraits are of three generations of Matthew Fordes.

An eighteenth-century gentleman's wash-stand, which has a multitude of functions. As well as the mirror for shaving, there is a green baize-covered 'drawer' that, when pulled out, becomes a writing table. Alarmingly spindly legs fold down to support the user of the pot. This gracefully hoodwinking piece of furniture is at Biddick Hall in County Durham, where it stands surrounded by eighteenth-century drapes and loomed over by an immense domed French 'Polonaise' bed.

This enchanting rose-decked bourdalou (opposite) from Newby Hall in Yorkshire also dates from the eighteenth century. It is thought that the word bourdalou came from Père Bourdalou, the Jesuit preacher in ordinary to Louis XIV of France. He was a preacher of exceptional ability whom ladies flocked to hear; loth to miss a minute of his brilliance, they took these useful containers with them, hidden away in their muffs.

A painted wood box 'close-stool', or 'portable closet', of 1825, with brass handles and a hinged mahogany seat board, at W. Stockbridge and Sons in Cambridge.

A mahogany pot cupboard with gilt bronze wreath of about 1810 (opposite), at Belvoir Castle in Leicestershire. The vast early eighteenth-century bed, its contemporary hangings and the Chinese silk on the walls all came from the old Belvoir (it was rebuilt in the early 1830s).

Two sets of bedside steps – designed to gracefully conceal chamber-pots – at Alnwick Castle in Northumberland. The pot was kept in a compartment behind the top step, the middle step being pulled out to sit on. The gilt canopied bed (opposite) is late eighteenth century. Two of the portraits are of the second Duke of Northumberland; the others are of Maria, Countess of Coventry, and Mr Duters, tutor to Lord Algernon Percy.

A gleaming lustreware 'Frog' pot, so-called because of the little lumpen creature creeping insanitarily up its rim. They originally appeared on mugs, to amuse drinkers with the glugging noise of the liquid flowing though the frog's body. Not an appealing feature on a chamber-pot.

These were all 'Sunderland Ware' – made at Sunderland, in County Durham – which enjoyed its commercial heyday in the 1840s, when there were as many as twenty-four factories producing the china. They had a particularly pious speciality: religious tracts gilded and gleaming on china plaques – a far cry from the elegant sentiments that decorate these pots.

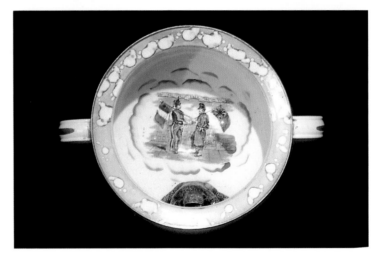

'New Marble in Silver Grey', a wash-basin of the late nineteenth century made by Edward Johns and Co. Ltd. The marbling effect smothered many a sanitarian's creation in the nineteenth century. This basin, now in the Gladstone Pottery Museum in Stoke-on-Trent, was the only surviving example I could find of the once popular 'Silver Grey'.

A wash-basin in 'New Marble in Rouge Royal' (opposite). It was made in 1896, by Edward Johns and Co. Ltd. It is now in the Armitage Shanks showroom at Armitage in Staffordshire. There were three 'marbling' effects produced by potters in the late nineteenth century. This, as well as the shiny black 'St Anne's', and 'Silver Grey'.

A wash-stand designed by the extravagant medievalist William Burges for his rooms in Buckingham Street, London. It was this very piece of furniture that John Betjeman gave to Evelyn Waugh in 1953 and about which Waugh wrote worriedly in his diaries a year later, as well as bringing it into *The Ordeal of Gilbert Pinfold*. He had gone to London to collect his glorious gift, when he 'saw and handled a serpentine bronze pipe which led from the dragon's head in the tank to the bowl below'. When the gift arrived, Waugh found to his horror that this central feature, 'the climax of the design', was missing. Betjeman and the delivery men all insisted that it had never existed and that he had suffered a 'complete delusion'. Waugh makes this the first of many such delusions for Pinfold, with an excellent description of the wash-stand: 'A most remarkable piece, a wash-stand of the greatest elaboration, designed by an English architect of the 1860s, a man not universally renowned but of magisterial status to Mr Pinfold and his friends. This massive freak of fancy was decorated with metalwork and mosaic and with a series of panels painted in his hot youth by a rather preposterous artist of the Royal Academy. It was just such a trophy that Mr Pinfold valued.'

Beneath the basin there is a

waste tank, hidden behind carved and massively hinged doors. The marble bowl, inset with silver fish, is surrounded by ancient Oriental bronzes – a stork, a figure and a gold spotted ram – which operate the plug and the taps. The bronze shaving bowl is embellished with a silver frog. Paintings of Narcissus, thought to be by Edmund Poynter, glow out beneath the castellations, under gilded words from Chaucer:

> THIS:IS:THE:MIRROUR:
> PERRILUS:IN:WHICH:THE:
> PROUDE:NARCISSUS:SEY:
> ALL:HIS:FAIR:FACE:BRIGHT:

There is a little dragon-like head that spouts water into the bowl below from the two lead tanks behind the frieze – one for hot water, the other for cold.

The Marquis of Bute's wash-basin (opposite), designed by William Burges in the 1860s, in the Bachelor Bedroom in the Clock Tower of Cardiff Castle. Both Bute, the patron, and Burges, the architect, were ardent medievalists; they set about restoring and rebuilding the castle between 1868 and 1872. As always with his wash-stands, Burges created an outlandish piece of furniture, this time set a-shining with fragments of marble, as well as mosaics and mirrors and cast-ironwork galore.

The Ladies 'Retiring Room' at Harrods in London (opposite), which was built in 1908 as part of a suite of sumptuous rooms for the Ladies Club. The cloakroom was proudly proclaimed in The Diamond Jubilee Catalogue; under the general heading 'An English Institution that commands the admiration of the World', the cloakroom was described as 'nicely fitted out . . . with the effect of the glass being admirable. The windows are of stained Cathedral glass, while the walls are covered throughout with panelled Brecchi Sanguine, Pavannazi, Levantine marble and onyx panels. The whole rendering an unique effect . . . ' The current wash-basins are in fact new – and excellent – fibreglass replacements of the originals. A remarkable thing to know about Harrods is that most of its water comes from three private artesian wells, which are 640 feet deep beneath Knightsbridge – among the deepest in London. In an act of dastardly desecration this room was destroyed to make way for escalators in 1980.

Lady Bute's wash-stand, designed by John Chapple, in one of the rooms built by William Burges at Castell Coch, in Wales. Lord Bute and Burges were already thundering along with their elaborate rebuilding of Cardiff Castle when they started on their schemes for the thirteenth-century ruins of Castell Coch – also owned by Lord Bute – some 5 miles away. Its mock medieval exterior had been finished, but its interior only just begun, when Burges died in 1881 – a miserable loss as he was only fifty-four years old. He had made a model of this room, so we know that he was responsible for the domed ceiling, painted with monkeys luxuriating among pomegranates and vines. The turretted wash-stand was fully functional; both of the towers are lead lined, one for hot water, the other for cold (after it had been hauled, in buckets, up innumerable steep steps!). The fish tap, emblazoned with a coronet and the letter B, spouts water on to the china fish that swim round in the china basin.

The ornate 1890s Gentlemen's Cloakroom of the Manchester Club, originally the Reform Club, in King Street, Manchester. Marble panels in the ceiling reflect the octagonal marble-topped wash-basins. The white wooden boxes are for towels. There is intricately carved woodwork throughout the place, the pillars quite smothered with masks and ribbons, fruits and flowers. On the wash-basin at the end of the room there is a curious pump-like pipe, on which is written: 'This water is delivered absolutely pure by The Berkeld Filter.' To the right of the basin is a marble partition, specially installed to give privacy to those who wanted to clean their false teeth. At the top of the club's ornate wooden staircase are a splendid dining room and a billiard room, both richly deserving of their handsome cloakroom. The building has now been empty for ten years. It is occasionally heated and awaits a new use. Proposals have just been put in for office development but, as this is a much-loved and listed building, the city of Manchester will surely steer the Manchester Club safely into a new role worthy of its pedigree.

Edward VII's urinal at Wolferton Royal Railway Station (opposite) – now the Wolferton Station Museum – which was built by the Great Eastern Railway on the Sandringham Estate in Norfolk. Between 1898 and 1907 the station-master recorded every visitor in his diary, and it would seem that all the crown heads of Europe came to Wolferton, many of them no doubt pausing at the urinal on the way. The German Emperor was here in 1899 and 1902, King Carlos of Portugal passed through in 1902, the King of the Hellenes in 1905 and the King of Spain in 1907. The basin, a Jennings 'Club Pattern', was specially decorated with dark blue and gold lines for the Prince of Wales (before he became King). For 37 shillings (£1.85p) you could buy the singular 'Treadle Action Supply Apparatus' to go with this urinal – a small galvanized iron plinth decorated tiles, on which you stood, the apparatus 'flushing a continuous flush to the urinal basin whilst in use'.

One of the many WCs disguised as wardrobes – tiny, fully furnished rooms with a valve closet behind each 'cupboard' door – that were ordered by the Duke of Wellington in 1841 for Stratfield Saye in Berkshire. The first Duke was a great sanitary innovator and by the early 1820s he was installing steam baths, as well as hot water and a central heating system with great radiators like ships' engines, which kept the passages at an even 64° Fahrenheit.

In 1823 the servants had been asked not to empty their pots into the gutters at the top of the house, so terrible had been the stench, and the Duke suggested the innovation of a water-closet. These much-needed conveniences, however, were not to appear until 1841. Eight were put in that year, and the chief architect, Benjamin Dean Wyatt, seems to have been delighted: 'The Canadian wood (which they call "Rock Marble") turns out to be the handsomest wood for such a purpose, that I ever saw in my life . . . and I rather believe it is the only example of it in England.' The massive double doors were ingeniously designed with a mechanism that ensured that they could not be opened or closed separately, lest someone was shut in by the outer door. The WCs were valve closets, and all the china bowls were of the same

blue and white pattern, of butterflies, fuchsias, cherries and grapes. They were properly plumbed in, rather than being discharged into the rainwater pipes, as would usually have been the case at such an early date. Some of the bedrooms have wardrobes to match. You open one cupboard on to a row of coathangers and open the other on to a gleaming wood-encased valve closet, in a furnished and picture-hung little room.

The Ladies Club cloakroom at Harrods in London (opposite), with walnut and marquetry doors marching down the full length of this extraordinary, elaborate 'Retiring Room'. It was built in 1908 as part of the Ladies Club and was hailed by the press. 'Messrs Harrods have a way of anticipating the wants of their clients almost before they are aware of them themselves,' observed one ladies' journal. 'The fitting up of the luxuriously appointed club room is an example of this.' The report went on to describe the adjoining sitting room: ' . . . furnished in Adams style in figured satinwood. The chairs are upholstered in green corded silk, tastefully decorated with appliqué embroidery . . . Altogether it is a very sumptuous apartment with an air of welcome prevailing.' With monstrous insensitivity this was destroyed in 1980.

The Gentlemen's Public
Lavatory at Market Place,
Hull, built between 1901 and
1906 beneath the great gilded
equestrian King William III. Two
exquisite watercolours of these
urinals were submitted to the
contractor in September 1901,
one with these flamboyant
Corinthian columns, the other in
the more severe Ionic style. The
doors were to be of 'well sea-
soned pitch pine', the tiles of 'first
quality chocolate enamelled'.
B. Finch and Co. Sanitary
Engineers of Lambeth supplied all
the fittings, with eleven 'marbled'
slate urinal stalls and two delightful cisterns that were embellished
with tiny slate Tuscan columns.
There are also six Finch's syphonic closets with mahogany seats.
The total building costs came to
£1,129, of which the sleek
ceramic work came to a mere
£90. In the early 1990s £10,000
was spent on restoration, during
which worn tiles were sympathetically replaced. It has been given a
Grade Two listing.

The ultimate in urinals
(opposite)– the twenty
Twyford 'St Anne's Marble' stalls
surrounding the magnificent
'Hexagonal Urinal Range', on
the pier at Rothesay on the island
of Bute. After years of anxiety
that these jewels in the sanitarian's
crown might disappear, they are
now out of danger. There had
been a threat of demolition, with
grim forebodings of their rare
beauty being replaced by Formica
and the frighteningly functional
'Super Loo'. The Strathclyde
Building Preservation Trust
commissioned a feasibility study,
which proved that, at little extra
expense, the stalls could be
restored rather than razed, and the
case was won. The small glazed
brick building is the first and the
last that you see when sailing in
and out of Rothesay harbour, and
it is a great relief to know that
this little beacon of beauty is now
snugly secure, rather than about
to be smashed to smithereens.
With their sumptuous splendour,
the urinals trumpet out all the
triumphs of nineteenth-century
taste and technology, and they are
to be found, glittering away, on an
island in Scotland. They were
reopened to the public in 1994
and are proudly polished daily.

The magnificent faience and tilework of the Gentlemen's Cloakroom (opposite) in what was the City Club (now the City of London Club) in Old Broad Street, London. It was in December 1901 that a design was submitted to create these shining vaults. They were to cost some £6,000, with £400 to be spent on the tiling of the walls as well as the ceilings and the pillars. Matthew Hall and Co. supplied the sanitary fittings, with wash-basins encased in grey marble, each proudly proclaiming itself as 'The Lavazonic'. As you sail through the neo-classical hallway of this London club – built 1833–4 by Phillip Hardwick – and on down the stairs towards the basement, you are quite unprepared for the shimmering sight that greets you.

One of the two ceramic arches by Willink and Thicknesse at the entrances to the underground Gents lavatories in Derby Square, Liverpool. They were created between 1902 and 1906 in conjunction with an immense monument to Queen Victoria, which stands directly above them. The thirteenth-century Derby Castle stood on this very site, but was demolished in 1720 and replaced by two consecutive churches, which in their turn were replaced by Queen Victoria, surrounded by Agriculture, Commerce, Education and Industry, with this handsome monument to Sanitation beneath her feet.

Twyford's 'Rouge Royal' urinals and cisterns, of the late 1890s, in the Gents in the Philharmonic public house, in Hope Street, Liverpool. The process of marbling by the potters was always the same: the patterns were transfer printed on to the biscuit (fired but unglazed pottery) – by ladies in white gloves plying linen pads – and the ware was then passed through a 'harding-on' kiln before glazing. These circular backed urinals could also be ordered in 'Sage Green' and 'Rich Brown', or otherwise with the marbling effect in 'Silver Grey' or 'St Anne's'.

A bee for men to aim at to avoid splashing their shoes (opposite), and to bring a scholarly smile to many lips, as the Latin for bee is *apis*. It decorates one of two 'Diamond' urinals in leadless glaze that were designed for the Wigan College of Technology (now the Town Hall) by Walter E. Mason of Harwich. Both urinals still survive in splendour.

Twyford produced a 'Bullseye Target' for the same purpose!

The Old Toll Bar of 1896, in Paisley Road West, Glasgow. To the right of the picture huge mirrors – etched and inlaid with gold – rise up from dado to ceiling. To the left the bar sweeps down the full length of the room beneath eight enormous barrels, 6 feet high. The whole pub is timber-lined and varnished black, with the bar and barrels, as well as the walls and the ceiling, all shining away in the gloom. There are two sets of double swing doors of stained glass. The barrels are framed by cornices and elaborately carved pillars, which were taken out of the old City of Glasgow Bank.

The door to the Gents (opposite) in the Horseshoe Bar in Drury Road, Glasgow. The glass and the mirror behind it are both surrounded by a wooden horseshoe, the pub's symbol, and there are more than seventy of them in this gleaming interior of the 1880s. There are two wooden and two marble fireplace surrounds, all in the shape of horseshoes, and an immense mahogany horseshoe makes up the bar. There are three mirrors with horseshoe surrounds and even more that are etched with horseshoes. The clock, instead of having numbers, has twelve letters, spelling the word horseshoe, around its face. It is, of course, surrounded by horseshoes.

Queen Victoria's water-closet and cloakroom in the Royal Compartment built in 1869 for the London and North Eastern Railway. It is part of the sumptuous suite designed by William Bore for the two 30-foot-long royal carriages that were joined by bellows – the first to be built in the British Isles but loathed by the Queen. Her carriages were cut off from the rest of the train, and with her dressers at one end, and her personal attendant at the other, Queen Victoria was entirely self-sufficient.

Her cloakroom has all the splendour of the saloon: both of them bulge with royal blue moiré silk walls, and with beige silken ceilings. There are tasselled steadying straps (each embroidered with shamrocks, roses and thistles), silken blinds and tasselled curtains. There is Gothic wood-work throughout. The saloon had two bucket chairs as well as two upright chairs and an enormous sofa, all upholstered in the same blue and buttoned silk. The lamp-shades are festooned with lace.

The wooden cornices are thick, carved and gilded, supporting gilded coats of arms. Splendour has been packed into this tiny coach, which is the very essence of the overblown opulence of nineteenth-century decoration. The cloakroom is very small, just the width of the lavatory seat. There is a valve closet, with a delicately painted key pattern in turquoise, and gold around the bowl.

The carriages are the prize exhibits of the National Railway Museum in York.

The cloakroom in the Royal Compartment of the London and North Eastern Railway that was created for John Brown, the dour Scots ghillie who became Queen Victoria's personal attendant in 1865. It is part of Bore's Royal Suite, which consisted of a drawing room, a bedroom and a cloakroom for the Queen, as well as a sitting room, a bedroom and a scullery-cum-cloakroom at either end for her dressers and her personal attendant. Both coaches groan with glory; it is as if all Victorian dreams of grandeur have been stuffed into these carriages. John Brown's is hardly less imposing than the Queen's; in his sitting room the ceiling and the chairs are padded with a beige material, woven with a silken diamond pattern. The walls are lined with the same smooth Gothic forms as the cloakroom.

This, like the Queen's, is tiny, with a closet at one end, a wash-basin at the other, and, between them, a folding table for silver tea-brewing instruments. Gothicry is everywhere, with arches marching around the walls, and another arch lying vertically around the cornice. Both the wash-basin and the lavatory bowl shine silver. After John Brown died this little room would have been used by the Indian Abdul Karim, who became the next personal attendant to the Queen.

It is said, and I have been told this by a reputable scholar, who saw the marriage certificate at Windsor, that Queen Victoria married John Brown in the cottage by the loch at Balmoral, with her ladies-in-waiting as witnesses.

Wonderous Me. what do I See

In London, in Patricia Rawlings's collection, a stately galleon sails off on another Mariner's Pot, with yet more mournful verse.

A Sunderland Ware Mariner's Pot (opposite) made at the Garrison Pottery in 1885.

At Newby Hall in Yorkshire there is a screaming figure on the inside of the pot. On the outside, a miserably weeping woman, with two children, waves at a departing ship. Under the heading 'A Sailor's Farewell' is written:

Sweet oh sweet is that
    sensation,
When two hearts in union
    meet,
But the pains of separation,
Tingles bitter with the sweet.

There is a mariner's compass on the other side.

'The Lambeth Patent Pedestal "Combination Closet" ' – a wash-out in 'Blue Magnolia' at Pownall Hall School, near Wilmslow in Cheshire. This was designed in two versions, one revealing its curious trap, like an elephant's trunk, the other all enclosed, in a magnificent curving lump of porcelain. Advertised as a 'water closet slop sink and urinal combined' with its elongated basin, it was made in five decorative designs. The others were 'Wild Rose', in a multitude of colours; 'Raised Acanthus', 'Old Ivory'; and most appealing sounding of all, 'Ruby Hispanic, picked out in gold'. At least two enclosed versions of 'Blue Magnolia' survive; one at Saltram House in Devon, the other at the Doulton showroom in Stoke-on-Trent.

Doulton's 'Lambeth Patent Pedestal "Combination Closet" ' in 'Raised Acanthus, Picked out in Blue' (opposite), photographed at Hordley, near Wootton in Oxfordshire, whither it had just been brought from an island in the Inner Hebrides. Both versions of this closet were 'Strongly recommended for Hospitals, Asylums, Public Institutions, Factories, Tenements, and Model Dwellings'. It was made in a mass of floral patterns, both strangely named and coloured: 'Wild Rose and Pencilled Blackthorne', huge great yellow and orange roses on black and white foliage.

The 'Empire', a wash-down closet made by Edward Johns (opposite), principally for the Canadian market. Its trap outlet is through the floor, and the pipe goes on down to the basement – a frost-proof foil against Canadian weather (an ordinary outside pipe would have collapsed). Named after the Empire Brass Manufacturing Company of Toronto – which imported thousands of Edward Johns's lavatories – this is handsome proof of Britain's sanitary prowess. It is at the Gladstone Pottery Museum, in Stoke-on-Trent.

The Queen of the Closets, the 'Unitas', in 'Raised Oak' – possibly the most beautiful design of all. Invented by Thomas Twyford in 1883, the 'Unitas' was one of the first all-ceramic pedestal wash-out closets, and had the enormous advantage of not being boxed in, which prevented filth accummulating. There were ten patterns to choose from, including 'Florentine' and 'Dresden', and marbled and floral designs such as 'Chrysanthemum', 'Begonia' and 'Dahlia'. This splendid 1888 basin is extremely rare, in that it was decorated both inside and out, with two separate designs. The 'Unitas' was so successful that, by 1901, Twyford could claim that it 'surpasses in sale and reputation all water closets of this type'.

This beauty is now given pride of place in the reception area of Caradon Bathrooms (formerly Twyford), at Alsager in Staffordshire. This and all the other sanitary delights from Caradons' showroom reproduced in this book were stolen in 1993. After such banner headlines as 'Police Having Nothing To Go On' and 'Police Hope To Flush Out Thieves', the pieces were eventually found, but as it could not be proved that they were the very bowls that had been stolen, Caradons had to buy the whole lot back.

A Jennings wash-down closet in the 'Morning Glory' pattern. This was taken from the Mayor's Parlour in Brighton Town Hall during building work, and is now safely preserved in the Royal Pavilion Art Gallery and Museum. The design could also be ordered in 'Mulberry Morning Glory', in dark blue on white.

A 'National' bowl, one of the first all earthenware wash-out water-closets (opposite), transfer printed with vines, corn and passion-flowers. Produced by Twyford, it won the 'Highest Awards' at the International Medical and Sanitary Exhibition at South Kensington in 1881; and by 1889 100,000 were in use. It is in the Caradon Bathrooms showroom.

A wash-out water-closet, now in the Science Museum in London, which was removed in 1971 from the now-demolished Redhill House in Edgware. This pattern – of books and tiles, with medallions nestling in flowers – was peculiarly unlike any other.

WASH OUT
PATENT.

The 'waterway', a wash-down water-closet (opposite) in Fielding Road, London. The patent for this pattern – with branches of daisies printed inside the bowl as well as around the pedestal – was taken out in 1895 by George Hudson of Hanley in Staffordshire.

The 'Pilaster' – with a pillar sliding smoothly into its anemone-strewn pedestal – gave no clues as to its origins. It was photographed in a private house in Bury St Edmunds.

The 'Oeneas' wash-down closet (opposite) in Bath Road, Chiswick. 'Typically suited for the London County Council,' said the Farringdon Works catalogue of 1896. It offered three other WCs for sale: 'The Latestas', 'The County Council Closet' and the 'Puritas', all of them 'Enamelled in colours'.

The 'Latestas', this time in 'Best Gold Extra', with writhing convolvulus, lilies and roses. It was produced by George B. Davis, who, after working for several years with George Jennings, became a considerable sanitary innovator in his own right, patenting urinals as well as closets and 'ventilating systems'. This 'Latestas', once part of the 'Billiard Room Suite' at Oulton Grange in Staffordshire, is now in the Gladstone Pottery Museum at Stoke-on-Trent.

The 'Latestas' wash-down closet of 1899 (opposite) in the Science Museum. Strangely, these fanciful horn-blowing cherubs were thought to be 'Specially suited for the London County Council'. The workings were designed by a Mr George B. Davis of the Westminster Sanitary Works, inventor of the first automatically ventilated 'Closet Apartment' and described as one of the greatest sanitary engineers of his day. The pedestal was potted by Shanks, which used the same design for its 'Albania' closet.

The handsome "Compactum" Combination Wash-Down Closet' with 'raised decorating'. It is part of the complete Edwardian bathroom that survives at Kinloch Castle on the island of Rhum, off the west coast of Scotland. Shanks invented this 'Combination' system in the 1880s. Under the heading 'A New Departure', its catalogue announces: 'This system supplies a long felt want, by combining the closet and the cistern in one compact piece.' The ferns could also be ordered picked out in blue, and there was a host of other floral finery to be found on the 'Compactum': from 'Orchid' and 'Carnation' to 'Chantry' (Dog Roses), 'Wolsey' (Red Campions) or 'Keele' (Clematis), as well as 'Hydrangea' and 'Morning Glory' all in their natural colours.

A Doulton's 'Improved Pedestal "Simplicitas Wash-Down Closets" '(opposite), one in plain stoneware with 'Acanthus Raised Decoration'. The price of the stoneware basin was almost doubled with this opulent addition to the design. There was a magnificent bonus to go with this pedestal: a matching fluted brown and blue pillar that soared up behind the seat, splaying out into the 'Paisley' cistern above, which was also in the same brown and blue hues. It came in a variety of floral patterns in porcelain, but stoneware was considered particularly suitable for asylums '. . .to meet the requirements of the commissioners in lunacy', this model could be ordered with a 'self-raising seat'.

The 'Panorama' design giving unexpected grandeur to a 'Hopper' closet – the cheapest and most humble of systems. Although 'universally condemned' in the nineteenth century, it soldiered on, as it cost so little to produce. Its surface was 'too exposed and too narrow, its water supply a mere trickle and [it] should be only considered for the most menial quarters.' This elegant version was in a private house in London. Another is to be found at the Gladstone Pottery Museum, Stoke-on-Trent, and yet another at the Kings Lynn Museum in Norfolk.

The Duke of Bedford's private closet (opposite) in his box – 'The Bedford Box' – at the Royal Opera House, Covent Garden. With wild aspirations of grandeur, Jennings has woven his name into the seemingly royal coat of arms. The lion and the unicorn stand rampant over 'George Jennings, Patentee, Hydraulic and Sanitary Engineer, Palace Wharf, London, Stangate, Lambeth'. Described in his day as an 'indefatigable engineer', Jennings did in fact have a Royal Patent, and the Prince Consort had presented him with the Medal of the Society of Arts in 1847. The closet is still in hard-working order today.

The 'Closet of the Century' made by George Jennings. A finely formed WC with a 'syphonic discharge mechanism', it boasted of having the advantages of both the wash-down and the valve closet – the force of water of the one and the quietness of the other. This design was Jennings's pride and joy, and won him the Grand Prix at Paris in 1900. He advertised his invention with great gusto, claiming that it possesed advantages shared by no other system. This smoothly curvaceous bowl is at Rousham in Oxfordshire, alongside a Jennings 'lift-up' wash-basin (opposite) – a bowl that you empty by swivelling it upside down. Five more 'Closets of the Century' survive: one is at Kingsbridge in Devon, another at the Armitage Shanks showrooms in Staffordshire, and a third at the Gladstone Pottery Museum, while two were installed for the Prince and Princess of Wales at Wolferton Royal Railway Station in Norfolk. These, too, survive, decorated with discreet lines of royal blue and gold.

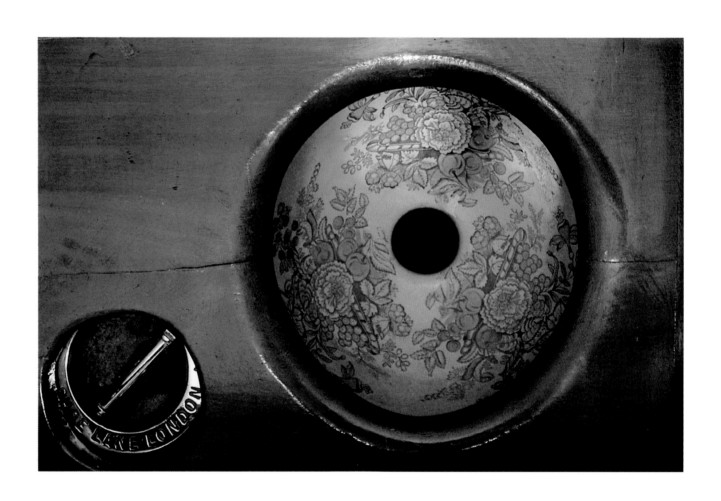

One of the two pan closets that are to be found in the woods near Seaforde House in County Down, Northern Ireland. They were for the use of the bachelor guests, who, never being allowed to smoke in the house, could puff away – clad in their velvet smoking jackets – as they strolled through the woods to these conveniences. There are two little fancy rooms, with cornices and moulded doorways, and each has a bowl decorated with these refined scenes of ladies dancing in front of an Italianate mansion. They have been magnificently restored.

A 'Best Quality' valve closet (opposite) made by H. Pontifex and Sons at its Farringdon Works in London. A curious extra could be bought with this closet in 1896: 'A Looking Glass Bottom Valve', which cost 5 shillings (25p). Pontifex would also engrave the customer's name and address on to the handle. This pattern – of cherries, begonias and peas – was to be found in many a bowl. There is one at the Bath Academy of Art, and another at Flintham Hall, near Newark. All the closets that Hellyer installed at Stratfield Saye, for the Duke of Wellington, are of the same design, as is the 'Optimus' that is in the Peeresses' 'Retiring Room' in the House of Lords. This bowl was photographed in a private house in Kent.

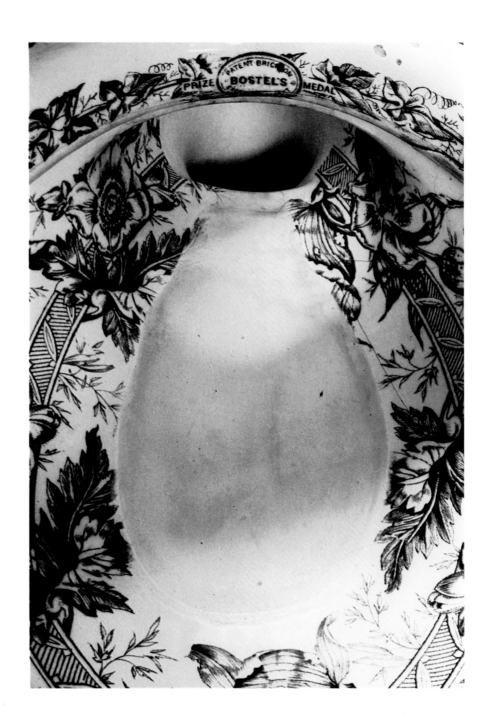

The 'Tubal' wash-out closet at the Armitage Shanks showroom in Staffordshire. In strong (and extremely heavy) fireclay, it is white enamelled within and buff glazed without. Its name came from John Shanks's brass foundry, The Tubal Works, which was named after Tubalcain, described in Genesis, Chapter 4, as an 'instructer of every artificer in brass and iron'! This ornamental acanthus leaf was also available in 'Turquoise and Gold' and 'Maroon and Gold'. The Tubal could also be bought decorated with honeysuckle, as well as chrysanthemums or 'Peach', – a mass of leaves – or with the fruit and flowers all together.

The first commercially successful wash-out closet, which was produced by Daniel Thomas Bostel of Bostel Brothers of Brighton (opposite), and exhibited by him in 1875. His grandson was alive when this photograph was taken and could remember ladies turning away in embarrassment at the sight of a lavatory on public display. They blushed with confusion at the sight of the gleaming object.

The 'Excelsior', a wash-down WC, at the Armitage Shanks showroom in Staffordshire. The pattern – with chrysanthemums, anemones and daisies – was patented in 1895.

The 'Beaufort' of 1884 (opposite), which was claimed by Humpherson and Co. of the Beaufort Works in Chelsea to be the 'Original Pedestal Wash-Down Closet'. It depended on the force of the flush immediately driving the contents of the bowl down through the trap. In other words, it was the forerunner of lavatories as we know them today.

The 'Sultan', a Shanks wash-out closet of 1896. This appears to have been an unsuccessful design, as it was listed for sale in numerous 1896 catalogues, but never again. A shimmering alternative could be ordered in 'Pearl'. 'Decorated', as illustrated, it cost £1 4s 9d (£1.24); in 'Marble' or 'Pearl' it was £1 3s 3d (£1.16).

The 'Pedestal Lion' (opposite) closet in a bathroom that I designed at the 1980 Ideal Home Exhibition for Adamsez. After years of searching for this engaging Emperor of all lavatories, I ran it to ground in Dorset. It was brought to London in triumph, and cheered by the crowds at Olympia, only to be borne off once again by its owner, into oblivion. The entire bathroom was fake. The bath and the wash-basin were lent by James Williams. There is a model of an 'Oeneas' WC on the mantelpiece. The wallpaper, called 'Jasmine', is a nineteenth-century pattern from Watts and Company.

A comparable closet to the 'Lion' once existed in the Durham Assize Courts: a key-pattered bowl on the back of a swan, which literally had to be mounted, by swinging your legs over its great neck.

Such decorative excesses had many critics; '...a humane man', wrote one, 'hesitates to use them for the purposes for which they are fixed, fearing to give the chamber maid a life-long labour in making them clean and whole-some again; for, with the utmost care in using such closets as uri-nals, it is impossible at all times to urinate into the basin so neatly that splashes, mishaps and drops ... shall not fall upon the edge of the basin and run down over the exterior parts ...' Floral and architectural patterns in deep relief cover the exterior of many closets; even recumbent lions, and swans in full sail, with closet basins on their backs, and other such incongruous combinations are being produced.

A Shanks wash-down of the 1890s, with birds flying through narcissi in the bowl. The same pattern appeared all over the unsuccessful 'Sultan', but this nameless jewel has been given a sober classical exterior.

Doulton's 'Improved Pedestal Simplicitas' (opposite) in 'Highly Ornamental Raised Decoration – Picked out in Blue'. This could also be ordered in white, and the same pedestal came in a wealth of other patterns in stoneware. At the time it was photographed, this was for sale.

'Not by Royal Appointment to The King' – an ingenious sales trick on Edward Johns's 'Dolphin' wash-out WC of 1909. Edmund Corm, who owned the firm by then, decided to add grandeur to its sanitary wares by emblazoning them with 'By Royal Appointment to The King'. As no such warrant had been granted, the Royal Warrant Holders' Association was enraged. By simply adding the word 'Not', Corm got round the difficulty, and preserved the bowl's distinction for a public who did not peer too closely. This dolphin was found in the attics of Armitage Shanks in the 1950s, having been forgotten for more than forty years. It had only been fired once and was therefore immediately fired again, which produced this curiously sleek and modern-looking creature. It is now in the showroom.

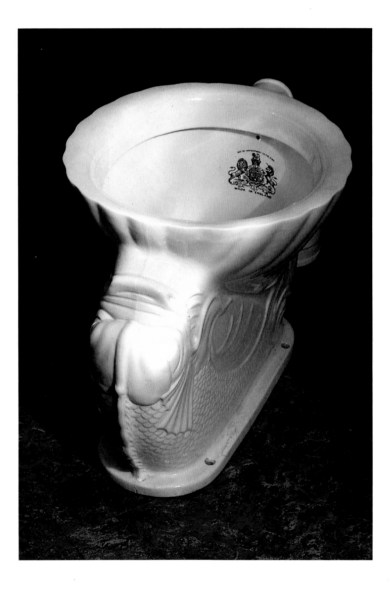

The 'Waterfall' wash-down WC (opposite), decorated with rope and roses, in a private house in St Peter's Square in west London.

Doulton's 'Improved Pedestal Simplicitas' wash-down WC, photographed at Stockport Cricket Club, from where it was about to be sold. This pattern, described as 'Highly Ornamental Raised Decoration', could be ordered in either plain white, or 'Picked out in Blue'. There were other patterns in stoneware.

The 'Deluge' (opposite), a wash-down water-closet in the 'Venetian' pattern at Raby Castle in County Durham. If you wanted either this one, or Doulton's two alternative wash-downs, called the 'Cardinal' and the 'Sirdar', there was a dazzling array of twenty-four patterns to choose from. As well as the usual floral and marbling designs, you could indulge in 'Japanesque', 'Dresden Print', 'Victoria' or 'Mikado'.

The 'Dolphin', a wash-out, the king of closets, but giving no clue as to who designed it. Its badge proclaims it to have been supplied by Stock Sons and Taylor of Birmingham, and it appeared in the 1882 catalogue issued by Bolding (which made all the works). With a 'syphon water-waste preventing flushing cistern' as well as a 'polished mahogany seat' and 'a pair of ornamental brackets' it would have cost you £6 10s (£6.50). In white and gold it was £7. It is thought to have been made in Derbyshire, but other than that nothing is known. It was safely housed in the Armitage Shanks showroom until, with sensational sadness, it fell apart in December 1994.

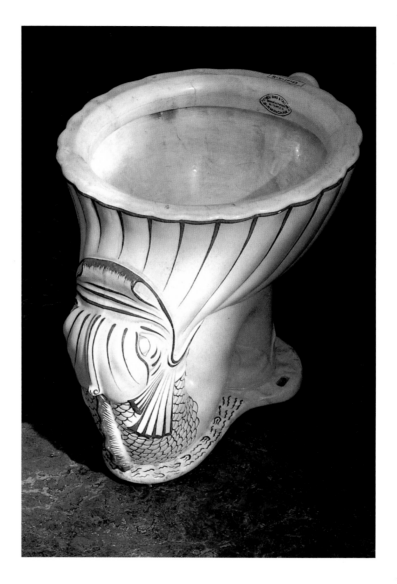

The 1840s Gothic Revival commode (opposite), with a comfortably cosy green felt seat, at Raby Castle in County Durham. It is part of the bed-room suite of two wash-stands, two commodes, a cheval mirror and a bedside cabinet, which belonged to the Duke of Cleveland.

D.C. – his monogram – and the ducal coronet can be seen on either side of the wash-stand drawers. The Victorian china set is Minton.

The bed, too, is 1840s Gothic Revival, lushly gilded and emblazoned with coats of arms. The Duke of Grafton and a Van Dyck judge survey the scene.

The pot cupboard, in what was the billiard room (now the drawing room) at Flintham Hall in Nottinghamshire. These discreet cupboards would have been essential in the eighteenth and nineteenth centuries for the gentlemen's instant relief, and they abounded in grand dining rooms, as well as billiard rooms, throughout the land. Flintham is a characterful Victorian pile if ever there was one, with, among its charms, a vast conservatory modelled on the Crystal Palace. In the drawing room Apollo stands over the golden beehive French work-box, and a nineteenth-century ostrich egg is held aloft by a gilded figure. The porcelain Chinese pagoda dates from the eighteenth century; the bamboo decorated chamber pot from the nineteenth.

A transfer-printed pot of the 1850s (opposite), made in Yorkshire. There are four pictures beneath the inside rim: 'Going to the Races' has a flower-bedecked cart, a fiddle, a dog and a bottle; 'Love and Beauty' shows a small white man embracing an immense black woman; 'The Repentance' shows a man and a woman, both in nightcaps and shirts, clutching one another; 'Married Life' is illustrated with a weedy-looking man being kicked and beaten by his poker-wielding wife.

A nineteenth-century Yorkshire 'Peasant Ware' pot, with nine handles. In lead-glazed earthenware, it has 'Hand It Over To Me My Dear' incised around the base.

A Leeds Ware political pot, with Prime Minister Gladstone (opposite), on whom to vent your spleen, in the bottom of the bowl. He was also depicted in chamber-pots in Northern Ireland, as a protest against his espousal of Home Rule.

A nineteenth-century 'Barge Ware' pot. These receptacles were made for the bargees, and were always lavishly – with unavoidably repellent consequences – smothered with encrusted decoration.

The chamber-pot at Stratfield Saye in Berkshire. When lifted up, this 'Patent Non Splash Thunder Bowl' plays music. On the side there is a hearty scene in a pub, with the words 'Oh Landlord Fill The Flowing Bowl'.

'Mulberry Chrysanthemum' decorating an 1899 Shanks 'Torrens' cistern (opposite), which was proclaimed at the time to be 'of high sanitary efficiency and beautiful appearance'. It was available in either mahogany or porcelain, with the usual breathtaking array of floral designs, all in their natural colours. 'The appearance of a closet is so often ruined', lamented the Shanks catalogue, 'by a rough looking overhead cistern with ill designed details.' This was photographed in a private house in Stirlingshire.

A Twyford 'Paper Box' of the early 1900s, made to fit on to the back of the seat of 'The Twycliffe Patent Syphon WC Basin'. Such fanciful extras were often made to match both the pedestals and the cisterns. This ornate design, called 'Corinthian', was sold with an ensemble that was handsomely decorated with royal blue and gold pillars. It is in the reception area of Caradon Bathrooms in Staffordshire.

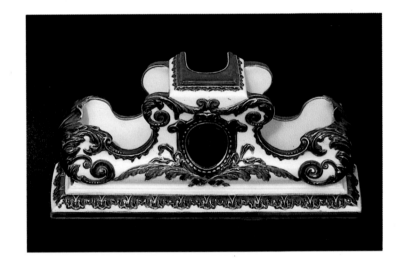

Three nineteenth-century bow bowls (opposite). The brown ribbon is the earliest, dating from 1830. The blue bow, from Sandringham, is mysteriously called 'Oriental Ivory' and bears a stamp of a Chinaman under an umbrella; it was potted in January 1881 by Bishops and Stoner of Hanley. The red ribbon was potted in September 1878 by Brownfield, at Corbridge in Northumberland.

The mid nineteenth-century 'Slipper' bath (opposite) at Wallington Hall in Northumberland. It was found surrounded by paraphernalia arranged by The National Trust, which now runs the house. Pink ribbons weave around a Minton wash-stand set, and there is a woollen jug cosy. 'Dr Nelson's Improved Inhaler' is one of three such contraptions, standing by a bottle of Aperient Mineral Water by Aquaperia of Harrogate. There are two brass water carriers and one of china, plus two boxes, one for a razor strap, the other for Sunlight soap. A pewter bedpan and a china slop bucket are on the floor, under an eighteenth-century pot cupboard. The bath is made from pieces of sheet metal, with a tap on the toe for draining and a funnel on the front for adding hot water. Spare a moment's thought for Marat, who was knifed to death in such a contraption; jammed immobile in a metal boot, unable to dodge the slashing blade!

'The Cameo Toilet Fixture' made by the British Perforated Paper Company, which was established in 1880. It is part of the complete Edwardian bathroom that survives at Kinloch Castle on the island of Rhum. Fitted up by Shanks, the bathroom has remained remarkably intact, with its great bath, its WC and its wash-basin. It is all of a perfect piece, right down to this contribution from the British Perforated Paper Company.

Doulton's 'Special Closet Seat' at West Dean in Sussex. A handsome disguise for a WC, this could also be bought in either oiled teak, polished mahogany or walnut. There were alarming extras called 'Seat Action Closets': one that kept the water flushing continuously when you were sitting on the seat; another that automatically flushed, the instant you rose to your feet. The glittering tiles were all part and parcel of the extravagant redecoration of the house, in 1891, for its American owners Mr and Mrs William James.

Another rare survival at Kinloch Castle (opposite): a nineteenth-century WC, complete with matching cistern and even its original fancy pipe brackets. It is the Shanks patent '"Levern" Single Trap Syphonic', which luckily – to ensure that its complicated syphonic system worked perfectly – could only be bought in combination with the cistern. This set was made in 'Mulberry Chrysanthemum' and would have cost £9 6s (£9.30).

Bolding's 'Pillar Pedestal Closet', made of cast iron and enamelled with porcelain, in the quartz marble cloakroom of 8 Addison Road in London. The house was built in 1906 by Halsey Riccardo for Sir Ernest Debenham of Debenham and Freebodys (now Debenhams). With tiles by William de Morgan, as well as mosaics by Gaetano Meo, this great Edwardian pile glistens from top to toe, both inside and out.

Lustreware tiles by William de Morgan (opposite) in one of the many bathrooms decorated by him at 8 Addison Road in London. The house has a spectacular array of such ceramic work, including fireplaces that are little tiled buildings in themselves; every bathroom is lined throughout with de Morgan's creatures, many of them mythical, which hark back to the medieval glories of the grotesque. He was the first to make tiles in lustreware, so-called because of its magical metallic sheen. Curious creatures leap over William de Morgan's tiles in the bathrooms of 8 Addison Road in London, a house that was lavishly decorated throughout with mosaic and ceramic work. Even the lavatories are encrusted with work by de Morgan, who had been entranced by fantastical animals from babyhood. When he was only two years old he had insisted that he was a 'three-toed woodpecker' and that his mother was a 'silky starling'.

A tale of our times; this remarkable scene was once part of the Victorian additions to the Treatment Centre in Bath (opposite), built in 1791 in Baldwin Street. In 1890 the grand Pump Room Hotel was integrated into the centre and in the 1890s all the treatment rooms were revamped with a quantity of turquoise quartz tiles. In 1986 the whole bang lot was demolished to make way for a second-rate shopping centre, which in turn was demolished in 1991. The site is now a derelict dump. These extraordinary and extreme changes of fortune have all gone on behind the original facade of 1791, ever serene and sympathetically smiling on the streets of Bath.

'The Acme of Luxurious Bathing': a Shanks 'Eureka' bath in a 'Superior Bath Cabinet' at Kinloch Castle on the island of Rhum. Kinloch was built in 1901, with many a technological triumph and all to the tune of one million pounds. A fully automatic orchestra was built under the stairs, and two of these baths were installed, sources of great wonder in their day. Standing beneath the canopy you are engulfed by terrifying aquabatics. 'Douche' is a great dagger-like stream, which feels like an apple corer boring through your head, whereas 'Spray' surrounds you with some 200 sharp needles of water. 'Shower' is straightforward, but 'Jet' is like a ramrod up your privates, and 'Sitz' is a gentle fountain for the same delicate area! 'Plunge' fills the bath – gushing out water from a slit, at knee-level – and then there is 'Wave', which sends a flattened jet of water from one side of the room to the other!

You could order your cabinet in either panelled oak, mahogany or walnut, and it could be inset with mirrors or otherwise carved. Swags were suggested as suitable decoration, as well as urns or crowns, and there were patterns of carved pomegranates, feathers or shells. One elaborate example looked for all the world like a miniature town hall.

'Shanks Patent Independent Plunge, Spray and Shower Bath' of 1896 in Withington in Manchester. The great boast of this bath was that there was 'an entire absence of wood enclosure', allowing no possibility of 'filth' collecting behind it. The inside was enamelled, the outside was japanned, and there were fourteen patterns to chose from, with marbling as well as stencilling and swags, or otherwise 'masks' and woodgrain. One, most delightful of all, had fat fish swimming beneath water lilies, with blue birds flying above them, between clumps of bullrushes, with reeds and water anemones.

Sadly this room is no more, as the whole house was auctioned away in the late 1970s. The bathroom had not changed one jot – Anaglypta walls and all – since it was fitted out in the 1890s: the 'Delta' wash-basin was still fixed into its cast-iron stand, with a mirror surround that soared 6 feet high, and the copper towel-rail had the curious device of a bowl for holding sponges.

Shanks' 'Fin de Siècle' cast-iron bath of 1899 (opposite). It was made with sixteen different patterns, including 'Vine', with swirling bunches of outsize purple grapes, and 'Scroll', with dolphins splashing through scroll-like waves and spray. Armitage Shanks has this in its showrooms at Stoke-on-Trent. It took six men to carry its $5^{1}/_{2}$ hundredweight.

The 1885 bathroom of Gledhow Hall in Leeds (opposite), with its mountainous terrain of gleaming Burmantoft's faience. It was built in 1885 for James Kitson, an affluent industrialist and politician who aggrandized the house so that he might entertain in it – day in and day out – for six months of every year. Gladstone was a constant visitor and no doubt relished this bathroom, as indeed must every Liberal minister of the period, as they all stayed in the house, at various times, until Kitson's death in 1911.

Another piece of its past to conjure up is the bathroom being used by 'The Little Owl's Society' – a Leeds ladies discussion group, of which Kitson's daughter Emily was a member. They often met at Gledhow, discussing such papers as 'Are modern manners improving?' or 'The immorality of gloomy views'.

The hooded 'Keyhole' bath, surrounded with tiles by William de Morgan, at Pownall Hall School in Cheshire, a house that a Mr Henry Boddington encrusted with the decorative delights of the Arts and Crafts movement. Rooms glow with the light of stained-glass windows, many of them by William Morris, and there is a wealth of ceramics and fanciful metalwork, as well as murals and carvings galore. The tiles in the bathroom, with their delicate repeating pattern, would have been done entirely by hand. The bath has four fittings: 'Douche', 'Shower', 'Plunge' and 'Spray'.

A Gothic swimming pool (opposite) amid the Gothic glories of Mount Stuart, the great pink sandstone house on the island of Bute built by the Marquis of Bute in 1900. Leaving the great Gothic hall behind you and descending a stone spiral staircase, you come upon it, with the reflection of the water rippling on the glazed vaulting above. Plunge in, and you are undoubtedly swimming up the aisle of a parish church!

Lord Bute's bathroom of 1873 at Cardiff Castle, designed by the scholarly and eccentric William Burges for his scholarly and eccentric patron, the Marquis of Bute. It is in the Bute Tower, one of the four towers, that give the castle the most sensational of skylines. The bathroom has sixty panels of polished marble – all named – that are inset into the walls of Central American mahogany. Lord Bute's bedroom is separated from the bathroom by an openwork screen, made up of two gilded Islamic arches – the same as the arch at the end of the room. The wash-basin is an enchanting 'tip-up' with a mermaid curled up in the bottom of the bowl. Creatures swim through arches around her. Below the rim are the words:

WHO:WOULD:BE:A: MERMAID:FAIR:SINGING: ALONE:COMBING:HER: HAIR:UNDER:THE:SEA:IN:A: GOLDEN:CURL:WITH:A: COMB:OF:PEARL:

The bathing paraphernalia in the magical Motto Room at Newby Hall, in Yorkshire. It was painted throughout with French mottoes, in 1857, by Lady Mary Vyner. Newby is a house of startling contrasts: as well as boasting a classical sculpture gallery it has an immense nineteenth-century billiard room – as violently Victorian as any in the land.

The red marble 'Roman' bath at Port Lympne (opposite) in Kent. The house was built for Sir Philip Sassoon in 1912. It was criticized as having had an 'overdose of magnificence' in the 1920s, with such additions as this bath, as well as a swimming pool in a neo-Roman garden.

Metallic sea creatures swim through the marble – with a starfish for a plug hole – of Lord Bute's 'Roman' bath at Cardiff Castle (opposite). They are part of the lively schemes of the architect William Burges and his patron Lord Bute for the medievalization of Cardiff Castle. Lord Bute was astonishingly only eighteen years old when they began their romantic work together. Both were learned and idiosyncratic Gothicists, and together they created an interior of staggering intricacy, magic and splendour. The bathroom is just big enough for its great marble bath – you have to climb into it from the doorway. The walls are of alabaster, laid in square and oblong blocks up to shoulder level, when they become all shapes and sizes. Blocks surrounding the Gothic windows and doorway are painted red and gold at their joints.

The 'Eridos', produced by John Bolding, at Castle Drogo in Devon. Built between 1910 and 1930 for Julius Drewe, who founded the Home and Colonial Stores. The house – looming up on a granite outcrop – is of sheer grey granite, both inside and out. Many of the walls are 6 feet thick, and many go off at unexpected angles, creating the strangest spacial effects. Every inch is a solid inch, from the granite vaulted corridors and the domed stone staircase, to the granite-and-oak-clad lavatories, each one designed with utmost care by Sir Edwin Lutyens.

Lord Curzon's bath at
Montacute, the sixteenth-
century house in Somerset that
he rented from 1915 until his
death in 1925. The magnificent
and imperious Lord Curzon,
Viceroy of India and holder of
innumerable distinguished
positions – who, when giving
speeches, was likened by
Labouchere to 'a divinity
addressing black beetles' – was in
constant and irksome pain from
the age of nineteen until the day
he died. He had curvature of the
spine, and always wore a steel
corset, which is very likely why
with his lady friend at Montacute
Eleanor Glyn, he built this
conveniently private bath in his
bedroom. He had decorated many
rooms throughout the house.
It was from here that in 1923
Curzon, then Foreign Secretary,
was summoned to London by
George v's secretary. Assuming
that he was to be made Prime
Minster, he had left Montacute in
triumph – no doubt having
joyfully prepared himself in this
'Jacobethan' cupboard – only to
find that Stanley Baldwin had
been given the job that he so
coveted.

The salt-and fresh-water bath
that was installed in 1910 for
Lady Astor in Rest Harrow
(opposite), in Sandwich Bay in
Kent. Shanks had provided the
bath, with a so-called 'extra inlet'
manufactured for salt, hard or soft
water. The sea water was piped
into the house, where it was
raised, by a two-cylinder pump,
into storage tanks. One, heated by
a boiler, supplied hot salt water;
the other cold. There are taps on
all the wash-basins for the same
supply.

The bulging 'Radio' Adamsez urinals in the Gentlemen's Cloakroom of the old Derry and Toms building (now the Roof Gardens building) in Kensington High Street, London. Built by Bernard George in 1933, it was acclaimed as the most advanced store in Europe, with praise being heaped on the lighting. 'The ceilings seem to float overhead,' wrote one ecstatic architectural critic. The most startling sight of all is still to be found above these urinals, where there is an enormous roof garden: a country spot with fully matured plants and shrubs, not to mention lawns and even trees. To be reminded of where you are, by peering through an ornamental slit in the brick wall on to the roaring traffic below, is a very peculiar experience indeed.

The marble walls and stylish doors of the gentlemen's urinals and lavatories (opposite) in the basement of Harrods. They were part of a grand scheme in 1930, when a barber's shop and a waiting room, as well as a cloak-room and the Gents, were all designed and most luxuriously appointed, by the 'in-store' architect.

The barber's shop is still gleamingly intact, with its original lighting and with great red leather chairs surrounded by walls of black bakelite, all banded with steel. The waiting room was converted into the Green Man pub in November 1973. The cloakroom and the conveniences survive, complete with their walls still laid in the original and complex 'perspective' pattern of different marbles. Napoleon Tigre and Lunel Rubane were both quarried in Northen France and the Grey Carara marble came from Italy. The 'creamy granular stone' hailed from Southern Germany and Northern France. Everything is to be restored to peak perfection by Mr Mohammed Al Fayed, the owner of Harrods who has already transformed the grand old store. As was said when he took over, the building must have heaved a great sigh of relief.

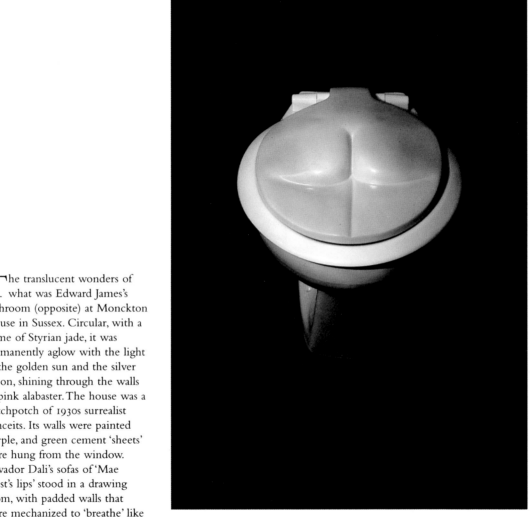

When I first recorded all these lavatories in the 1970s – providing proof of how dazzlingly Britain had once ruled the sanitary waves – this was the only design of the times I could find that had any vim or verve. It was cast in fibreglass by Martin Cook and Roger Stone, and it stood out as a lone beacon of adventure in those dim and dreary days of sanitary history.

The translucent wonders of what was Edward James's bathroom (opposite) at Monckton House in Sussex. Circular, with a dome of Styrian jade, it was permanently aglow with the light of the golden sun and the silver moon, shining through the walls of pink alabaster. The house was a hotchpotch of 1930s surrealist conceits. Its walls were painted purple, and green cement 'sheets' were hung from the window. Salvador Dali's sofas of 'Mae West's lips' stood in a drawing room, with padded walls that were mechanized to 'breathe' like a dog's stomach. There was a circular window in the front hall, through which, unawares, a guest could be seen – courtesy of a one-way mirror – in the bath. For good measure, fish swam between the guest and his secret audience.

The 'Priory', a gothic delight designed by James Williams in 1990 (opposite). This was one of the first beacons of hope – following hard on the heels of the era of the 'Avocado Suite' – for good new sanitary design. It comes with either a wooden or ceramic cistern – both of them gothic – and of course it should always have a wooden seat. There is also a Priory bidet and wash-basin in the same swooping gothicary. They can be bought at either Sitting Pretty in Dawes Road in Fulham or by contacting James Williams in Aston Rowant. They are manufactured by Chatsworth Bathrooms, which to my knowledge is the only firm left in the British Isles that still produces hand-made bathroom ware. Mussolini standards keep unexpected company with this particular WC. They were captured by James Williams' father, Lawrence (Bill) Williams (who was later to design the sets for the film *Brief Encounter*) when he cleared up Tripoli after World War II. Il Duce's great gilded symbols now stand proudly flanking the Priory, at Watlington in Oxfordshire.

The most delightful of all chamber-pots – the 'No. 1 Jerry' – with Hitler's face waiting to be defiled at the bottom of the bowl. When you lift it up, 'Rule Britannia' bursts triumphantly into tune. One part of a great collection of chamber-pots at the Oak Tree Inn in Hutton Magua in Yorkshire. It has now been borne off to New Zealand by the owner, Mrs Redshaw.

The marvellous modernity of John Young's bathroom in his apartment at Thames Wharf in London. It is in a glass tower all of its own, standing proud on the roof and cocking a modernist snook at the terracotta splendour of The Harrods furniture depository across the river. Young is part of the Richard Rogers partnership and echoes of that firm's buildings rebound throughout the flat, with its vast spaces and glass walls that seem to fling the rooms and the river together. It is a great celebration of industrial materials, fashioned throughout with such shining surfaces as stainless steel – which has been both polished and brushed – as well as glass floors, walls and ceilings, and vividly painted scarlet and yellow steel trusses. The original version of the bathroom tower was inspired by Frank Lloyd Wright's Johnson Wax building in Racine, Wisconsin, with its ceiling supported by great long stemmed, flat topped, mushroom-like forms. Young had planned to build his tower in Pyrex tubing, but when this proved too complicated he found himself in the bizarre situation of having his new glass scheme turned down by the local planning authority, even though it was his firm who designed the whole block! Eventually the tower was to be inspired by such structures as gasometers and water towers,

most particularly a water tower in Fresno, California. The idea for the stairs that sweep around the outside of the tower – taking you ever higher into the sky and on into a glass conservatory on the roof – was from a multitude of steps that wrap around an oil storage tank in Wyoming.

The roof of the bathroom is clear glass and the walls are of opaque glass bricks, criss crossed with web-like steel bracing. The sunken tub is of scented Japanese cedarwood and both the lavatory and the bidet, as well as the washbasin, are stainless steel, although sadly Scandinavian rather than British.

The shining example of how good modern sanitary design in Britain could be. This drinking fountain and two WC pans are all made by Santric of Swindon. Advertised as 'extremely resistant to impact damage', they are specifically made for public use as well as for prisons and hospitals and for the disabled. They are vandal proof; 'the steel cap boot cannot harm it', according to Stan Anniss, Santric's Director and General Manager, but he laments, 'Sadly there is nothing that can withstand the portable rechargeable drill.' 'The Pedestal WC Pan' on the left comes with either a high or low level cistern – both of polished stainless steel – while the 'Shrouded WC Pan' on the right is so named because of its pipes and pan being enclosed by a 'shroud'. The elegantly slender drinking fountain is described by Santric as having a 'full length shroud'. These swishly smooth stainless steel forms should be a challenge to the designers of British sanitary ware today.

The smooth and sheer bathroom designed by the architect John Pawson in 1993. The whole interior of the house – behind the facade of a Victorian terraced house in West London – is of the same monumental simplicity, with space and proportion, as well as light and shade, being of optimum importance. The floor drains the water away between the slabs, so that the bath may brim over, giving all the appearance of abandoned luxury in these starkly spartan surroundings.

A little building that is already a London landmark – the Ladies and Gents public lavatory designed by Piers Gough in 1993 (opposite) at Westbourne Grove in Notting Hill Gate. With a great glass canopy sweeping round above the stretched wedge of turquoise tiles, this is a triumph of old and new materials and forms. That it came into being at all is a miracle. A grim Portakabin-style lavatory had been standing on the site for over fifteen years when John Scott of the local Pembridge Association spotted a planning application sticking to its walls. He discovered that the local council were proposing to build a bland new box and suggested commissioning an architect himself. Piers Gough's enticing plans were some £10,000 more than the council could afford, so John Scott put his hand in his own, astonishingly generous, pocket. The schemes sailed through the council's planning committee, and with huzzas for such civic co-operation and daring originality, this beautiful little building was born. As an added bonus, it is bulging with blooms. A little shop was designed as an integral part of the building and, after dozens of applications, the florist 'Wild at Heart' was chosen as the ideal and colourful tenant.

Acknowledgements

I would like to thank all the following people for their kindness, which was often carried to astonishing lengths. It manifested itself in all manner of ways: patience, helpfulness, generosity, hospitality etc., and I am extremely grateful to each and every person mentioned.

Roger Acking for letting me have his lavatory seat for three weeks. Mr Frank Alflatt of the Science Museum for consistent kindness and cheerfulness, Mr Lesley Andrews, Mrs Daphne Archibald, Mr M Y Ashcroft MA County Archivist of North Yorkshire, John Aspinall, Lord Barnard, Mr and Mrs Jeremy Barr, Francesca Barran of the National Trust for helping time and time again with various problems. Mr and Mrs V T Berry, The Misses Matilda and Betty Besso who allowed their bathroom to be photographed for seven hours. John and Prudence Binning, Mrs Ruby Bloud, Mr S R Bostel, grandson of the first producer of the wash-out closet who went to an enormous amount of trouble over the course of a year to gather every scrap of information that he could for the book. Alison Briton whose beautifully designed tiles were sadly not finally included. Mr R T Buck of Dent and Hellyer Ltd, The Marquis of Bute for repeated and generous hospitality. Mrs Walter Caley, Mr Thomas Caunce, Ian Chadwell for hours and hours of his time, John Chesshyre, Dr J A Coiley of the National Railway Museum at York,

Mrs Robin Compton, Martin Cooke, The Lady Diana Cooper, Nicholas Cooper, Bobby Corbett, Dan Cruickshank for years of kindness and patience, Mr and the late Mrs Cotrell-Dormer, Geoff Dale of Doulton Sanitaryware Ltd, Warren Davis of the National Trust, Mr Ernest Delaney, The Duchess of Devonshire, John Dinkle, Keeper of Brighton Pavilion and his wife Camilla. Mr and Mrs Anthony Drewe, Martin Drury of the National Trust, Mr J Edwards of the Victoria and Albert Museum, Valerie Elliot, Stanley Ellis of Twyfords Ltd, Christopher Gibbs, Jim Gibson of Twyfords whose cheerfulness and help over the year was remarkable. Arthur Grogan, Wendy Gutteridge who wrote the book out twice in longhand. Robin Haddow who held at least fifty Ibrox Park football fans at bay from the stained glass gents in the Old Toll Bar Glasgow, as well as innumerable other kindnesses. Miss Hallam, Henry Harrod, my two boys Barnaby and Huckleberry Harrod, Peter and Gay Hartley, Cyril Hayson, Miles Hildyard and Mrs Sybil Hildyard, Sally Holden of the North Cornwall Museum and Gallery, Mr Holland, Mr and Mrs David Hudson, Miss Ilbert, Paul Keegan for his hours of valuable research, Mrs Issac, Katya Krausova, Mrs Elaine Lanchester, Librarian of West Dean College, The Marquis and Marchioness of Londonderry, Miss Frances Lovering of Doulton and Company, Jim Lowe, Angus McBean, Mr Murdo MacDonald, District Archivist of the

Argyll and Bute District Council, David McLaughlin, Mr Margrie, Tony Miles, David and Martha Mlinaric for wondrous friendship, David Moore, Mr J Morris, Mr Geoffrey Moorehouse, Mr D C Muir, Mrs J E Nelmes, Bob and Phyl Parker of the National Trust, Patrick and Judith Phillips, Mr Geoffrey Pidgeon, great grandson of the inventor of the wash-down closet. John and Rosalind Powell-Jones of the National Trust, Jiminy and Emily Read, Dr J R Reid, David Rhodes, Mr Ring of the Southport Cricket Club, John Ryan, The Hon Hugh Sackville-West, Mrs Salvin, Pauline Sargent of South Glamorgan County Council who was tireless in producing new information for the book. David Sekkers of the Gladstone Pottery Museum and his wife Simone, Houston Shaw-Stewart, Jeff Smith, Ian Smythe, Joan Stacey, The Rev Nicholas and the Hon Mrs Stacey, Elizabeth Steele, Chairman Stirling, Miss Magda Stirling, Mrs Susan Stirling, The Right Reverend The Lord Bishop of Southwell and Mrs Wakeling, Len Spiers, Mr Bertram Swainson of the National Trust, Dr Taylor, Emma and Toby Tennant, Clissold Tuely and Miles Tuely, Paul Tomasso, Mr E Tomkins, Jack and Jane Tressider, Mr John Tustin of Doulton and Company, Mr J Vallance of the Wigan College of Technology, Mr and Mrs Eric Walker, Mr Waugh of Armitage Shanks Ltd, Bron and Teresa Waugh, The Duke of Wellington, Lieutenant Commander P C Whitlock of HMS Victory,

Mr H G White, Tony Whitmore, Mrs Joan Wilson, Mr W G Wright, Mr and Mrs Gerald Yorke and John and Jeannie Yorke.

Mr Mohammed Al Fayed, the Chairman and saviour of Harrods, Harry Allison of Hull, Stan Annis of Santric, Andrew Barron for his friendly forbearance and dazzling design of this book. Stella Beddoe, Andrew Bellow, Robin Bloom, Richard Bostel of Bostel Brothers in Brighton and the great grandson of the inventor of the wash-out lavatory, Joanna Boyson of the National Trust for cheerful encouragement. Rob Close of Strathclyde Building Preservation Trust, Steven Clues, Curator of Bath Museum, Michael Cole, Director of Public Affairs at Harrods for his enthusiastic help and courtesy, Justine Fulford for help in every imaginable way and without whom this book would have stayed submerged in stagnant old sanitary lore. Piers Gough for building a landmark of London with his great public lavatory, Ivor Gwilliams of the Science Museum, Nadine Hanson, Harrod's archivist, ever ready with information on the history of the great old store. Mrs Hickson of Rest Harrow, Lindsay Johnson Laing, Rachel King, my editor, always ready at the other end of the telephone and always cheerfully encouraging and calm, however great the crisis. Angela Lee of the Gladstone Pottery Museum for her outstanding help, Warren Marshall, Tom Morgan at the Mary

Evans Picture Library, Alan Moss of Strathclyde Building Preservation Trust, Mr Olliver Wallington Hall, Venetia Paul at the Science Museum, John and Catherine Pawson who allowed me to photograph their bathroom while pouring water into the tub for hours on end. Hugh Pearman for alerting me to John Young's modernist masterpiece. Geoffrey Pigeon, my dear friend and the great nephew of the inventor of the wash-down WC who gave unflagging enthusiasm and support. John Shirley, Mrs Stevenson at Knole, William Stonor, Nikki Tibbles with her flower shop 'Wild at Heart' for her help and beautification of the Westbourne Grove Lavatory. Samantha von Daniken of the Water Monopoly for her help and sterling salvaging work, Colin Webb for commissioning the book and for reintroducing me to my old friend the lavatory, Peter Willasey, Harrods press officer, for spending two hours with me in the store's beautiful underground Gents. Portia Rifat of Santric of Swindon, Chris Rigley, Ian Rossinton, Rev J V H Russell of St Mary's Church, Selling, Vivienne Schuster of Curtis Brown for her stalwart support, John Scott whose great generosity in helping to build the perfect public lavatory has given pleasure and relief to thousands, Julien Seymour, James and Jan Williams for their unstinting help and advice, Henrietta Wood for her lumberjack-like help in photographing the modern bathrooms, Mr Woodberry for his stop press news of Armitage Shanks, Perry Worsthorne for his sweetheart support as he was swamped daily by sagas of sanitation, John Young for allowing me to photograph his magnificently modern bathroom.

The water closet (left) in the Turret Room at Osterley in Middlesex. The 'Plumber's Mate' in breeches walking over cobbles in the badge at the back of the bowl was the trademark of Dent and Hellyer.

BIBLIOGRAPHY

CATALOGUES AND MAGAZINES

Catalogues issued by all the companies mentioned in the introduction were consulted as well as the following magazines: *The Builder*, 1845–1850; *The Builders Magzine of Designs in Architecture*, 1774; *Illustrated London News*, 17 August 1850 (new shower bath); *Industries of the South Coast*, 1891; *The Metropolitan*, 15 July 1882, 'Manufactures of Public Utility'; *The Plumber*, 1 April 1922; *The Sanitary Record*, 23 April 1897, 'A pioneer in Sanitary Engineering'; *Speculum*, July 1934, 'Latrines and Cess-pools of Mediaeval London'; *The Surveyor and Municipal and County Engineer*, 11 Jan 1894, 'A new underground convenience'.

BOOKS

Allen, E. *Wash and Brush Up*. London: A & C Black, 1976.

Ashe, G. *The Tale of the Tub*. London: Newman Neame, 1950.

Barnard, Julian. *Victorian Ceramic Tiles*. London: Studio Vista, 1972.

Beckman, Johann. *A History of Inventions and Discoveries*. Translated by W Johnston. 4 vols 1971.

Bourke, Captain John G. *Scatalogic Rites of All Nations*. Washington DC: W H Lowdermilk, 1891.

Briggs, Asa. *Victorian Cities*. London: Penguin, 1963.

Chatwin, Bruce and Deyan Sudjic John Pawson. Barcelona 1992 *Editorial Gustavo Gili*, Barcelona 1992

Coomber, Matthew. 'Pee Super' *Building*, 30 July 1993.

Crunden, John. *Convenient amd Ornamental Architecture beginning with the Farm House and regularly ascending to the Most Grand and Magnificent Villa*. London 1767.

Donno, E S. *Sir John Harington's A New Discourse of a Stale Subject, called the Metamorphosis of Ajax*. London, 1962.

De Vries, L. *Victorian Inventions*. London: J Murray 1971

Dyos and Wolf. *The Victorian City*. London: Routledge, 1976.

Eassie, W. *Sanitary Arrangements for Dwellings*. London, 1874.

Eyles, Desmond. *Royal Doulton 1815–1965, The Rise and Expansion of the Royal Doulton Potteries*. London 1965.

Garland, Madge. 'The Town Houses of Halsey Ricards'. *Country Life*. 13 and 20 November 1975.

Girouard, Mark. *The Victorian Country House*. Oxford: Oxford University Press, 1971.

Harington, Sir John. *The Metamorphosis of Ajax*. 1596.

Hellyer, Bertram. *Under Eight Reigns. Dent and Hellyer*. London, 1930.

Hellyer, S Stevens. *The Plumber and Sanitary Houses*. London: Batsford,1877.

Hellyer S Stevens. *The Principles and Practices of Plumbing*. London: Bell, 1891.

John Boys Architects. *Feasibility Study on Public Conveniences*, Rothesay Caradon Twyfords, 1989

Lamb, H A J 'Sanitation: An Historical Survey'. *The Architects Journal*. 4 March 1937.

Loudon, John Claudius. *The Architectural Magazine and Journal of Improvement in Architecture Building and Furnishing and in the Various Arts and Trades connected therewith*. London: Longman, 1934.

Lucas, C. *An Essay on Waters*. London, 1756.

McNeil, Ian. *Joseph Bramah*. Newton Abbot: David and Charles, 1968.

Mayhew, Henry. *Mayhew's London, Selections from 'London Labour and the London Poor'*. Ed. P Quennell. London, 1949.

Middleton, G A T. *The Drainage of Town and Country Houses*. London,1903.

Moore, E C S. *Sanitary Engineering. A Practical Treatise*. London, 1898.

Newman, Harold. 'Bourdalous'. *The Connoisseur*. December 1970 and March 1971.

Palmer, Roy. *The Water Closet, A New History*. Newton Abbot: David and Charles, 1973.

Poore, Dr George Vivian. *The Dwelling House*. London: Longman, 1897.

Popham, Peter. *Royal Flush. Independent Magazine*, 3 July 1993.

Pudney, John. *The Smallest Room*. London: M Joseph, 1954.

Quennel, Marjorie and C B. *A History of Everyday Things in England*. London, 1933.

Read, Charles Handley. 'Notes on William Burgess's Painted Furniture'. *Burlington Magazine*. 1963.

Reyburn, Wallace. *Flushed with Pride, the Story of Thomas Crapper*. London: McDonald, 1969. Reprinted

Pavilion 1993.

Reynolds, Reginald. *Cleanliness and Godliness*. London: Allen and Unwin, 1943.

Robins, F W. *The Story of Water Supply*. Oxford: Oxford University Press, 1946.

Rolleston, Samuel. *Philosophical Dialogue Concerning Decency, to which is added a Critical and Historical Dissertation on Places of Retirement for Necessary Occasions, together with an Account of the Vessels and Utensils in Use among the Ancients, being a Lecture read before a Society of Learned Antiquaries*. 1971.

Routh, Jonathan. *The Good Loo Guide*.

Scott, G B *The Story of Baths and Bathing*. London: T Werner Laurie, 1939.

Sudjic, Deyan. *Apartment London Architect John Young, Blueprint Extra*, Wordsearch Ltd, London, 1991

Trevelyan, George Macauley. *Illustrated Social History in Four Volumes*. London: Pelican, 1964.

Walpole, Horace. *Selected Letters*. Ed. by M A Pink. Scholar's Library. London, 1938.

Walsh, John Henry. *A Manual of Domestic Economy suited to Families spending from £100 to £1000 a Year*. London, 1857.

Ware, Isaac. *A Complete Body of Architecture*. London, 1735

Webster, Thomas. *An Encyclopedia of Domestic Economy*. London: Longman, 1844.

Wright, L. *Clean and Decent*. London: Routledge, 1960.

INDEX